# WJEC

# GCSE

# WORKBOOK

# Geography

Support for WJEC GCSE Geography and
WJEC Eduqas GCSE (9–1) Geography A

Practise your exam skills • Answer questions confidently • Improve your grade

Andy Owen

HODDER
EDUCATION
LEARN MORE

# Contents

Throughout this book, you will find questions and guidance relevant to both the WJEC specification for Wales and the Eduqas specification for England. At some points however, you may need to be aware of different things for your exam depending on whether you are sitting the Eduqas exam or the WJEC exam. You will find boxes explaining this that look like this:

> **WJEC**
> This box shows information specifically for the WJEC exam in Wales.

> **Eduqas**
> This box shows information specifically for the Eduqas exam in England.

Chapter 4 is about your fieldwork assessment. This is assessed differently depending on whether you are studying the WJEC specification in Wales or the Eduqas specification in England.

For information in Chapter 4 that is only relevant to the WJEC specification, you will see a red bar running alongside the information on the page, like the one on the left of this paragraph.

For information in Chapter 4 that is only relevant to the Eduqas specification, you will see a green bar running alongside the information on the page, like the one on the left of this paragraph.

Anything in Chapter 4 that has no coloured bar alongside is relevant to **both** specifications.

# Introduction: What is assessed in each exam paper?

**Figure 1** shows what is assessed on each of the exam papers. There are some options in Paper 1 and Paper 2. Make sure you know which ones you covered.

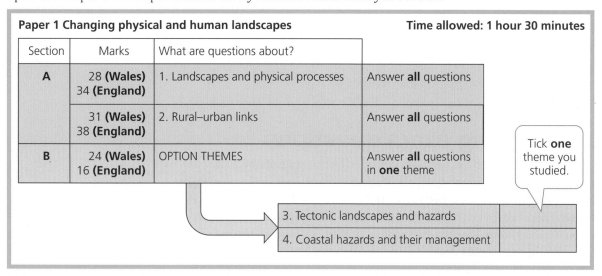

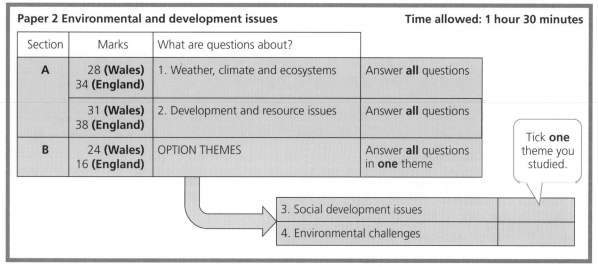

**Figure 1 What each exam paper assesses**

**WJEC**
In Wales, fieldwork is assessed in an open book test called the NEA (or Unit 3). This happens in the autumn term before you take your other geography exams. See pages 89–92 for some advice on questions about your own fieldwork.

**Eduqas**
In England, fieldwork is examined in Paper 3. For more details see Chapter 4.

# Chapter 1 How Geography is assessed in Papers 1 and 2

This chapter is about how GCSE Geography is assessed in Papers 1 and 2. It will cover:
- what the exam questions mean
- how to tackle questions that test your:
  - ability to use graphs and maps
  - knowledge by asking you to describe
  - understanding by asking you to give reasons or explain
  - ability to make a decision, evaluate or analyse evidence – including photos.

## Understanding exam questions

Papers 1 and 2 have a variety of questions designed to test your ability as a geographer. It's important you understand what each question is asking you to do:
- The **Assessment Objective (AO)** is what the examiner is looking for in your response. There are four AOs. They are described in **Figure 2**. Each exam question assesses **one** AO. This means that, in some questions, the examiner wants you to remember geographical facts so you need to describe in detail. In other questions the examiner wants you to evaluate. If so, you must evaluate – describing facts and figures would be a waste of time. You need to read each question carefully to understand what the examiner is looking for.
- **Command words** are words such as *describe* or *explain*. The command word tells you what you must do when you write your response. Common command words used in Paper 1 and Paper 2 are given in **Figure 3**.
- The **tariff** is the number of marks that are available for each question. These are shown at the end of the space where you put your answer. Use the number of lines printed on the exam paper as a guide to how much you should write.

> In Paper 1 and Paper 2 you have about one minute for each mark. Spend about ten minutes on an 8-mark question. **Don't** write a lot for a question worth 1 or 2 marks.

**Figure 2 The Assessment Objectives (AOs)**

| AO | What the examiner is looking for | Wales: marks in *each* paper | England: marks in *each* paper |
|---|---|---|---|
| Knowledge (K) | Your ability to remember geographical facts. | 15 | 18 |
| Understanding (U) | Whether you understand geographical concepts and processes. | 20 | 24 |
| Application (A) | Your ability to evaluate or use evidence to make a decision. | 25 | 24 |
| Skills (S) | Your skill when you use maps and graphs or make calculations. | 20 | 18 |

## Read the question carefully

It is essential to do what the command word asks you to do. **Figure 3** lists common command words and explains what they mean. The hardest questions in Papers 1 and 2 are worth 8 marks.

> If you understand the command word then you can tell which AO is being assessed. Study **Figure 3** and **do** what the command word tells you.

**WJEC**
In Wales, the 8-mark questions could assess understanding **or** application. There is **one** in each theme.

**Eduqas**
In England, the 8-mark questions always assess application. There is **one** of these questions at the **end** of each theme.

**Figure 3 Command words that could be used in Paper 1 and Paper 2**

| AO | Command word | What you need to do | Example question |
|---|---|---|---|
| K | Describe | Show your geographical knowledge by giving a brief account of something. | Describe the process of attrition. [3] |
| | Give | Make a short, simple statement. | Give one push factor. [1] |
| | Complete | Use words from a list to fill in the missing words in a passage. | Complete the sentences by selecting the correct word from the box below. Use each word only once. [3] |
| | Define | Give the meaning of a geographical term. | Define the term counter-urbanisation. Tick (✓) one box below. [1] |
| U | Give one reason | Make a point and then explain it. Use the connective 'so' to link your point and explanation. | Give one reason why tropical regions have high temperatures throughout the year. [2]<br>Give two reasons why people migrate to global cities. [4] |
| | Explain | Show your understanding by giving reasons. A great answer develops a chain of reasoning. | Explain why the informal economy is good for people and the economy of cities in LICs. [6] |
| A | Suggest | Propose a possible solution, reason, or consequence. Your suggestion should be based on geographical evidence. | Suggest the economic impacts of this hurricane. Use evidence from the photograph. [4] |
| | Evaluate | Consider strengths and weaknesses. Make sure you do both so your answer has some balance. | Evaluate one strategy for reducing the risk of flooding. Use evidence from the Fact Box. [8] |
| | Analyse | Examine geographical evidence carefully to find and explain connections or patterns. | Analyse the factors that increase the risk of flooding in this drainage basin. Use evidence from the OS map. [8] |
| | To what extent | Make a judgement by weighing up the arguments for and against. Make sure you give reasons for your decision. | All new housing in the UK should be built on brownfield sites. To what extent do you agree with the statement? Make use of evidence from the photograph and Fact Box. [8] |
| S | Calculate | Work out the value of something. | Study the table below.<br>Calculate the mean of the values. Show your working. [2] |
| | Describe | Give a brief account of a pattern on a map or a trend on a graph. | Study the map below.<br>Describe the location of Mumbai. [2] |
| | Give | Make a simple point. | Give two ways in which the graph could be adapted to make it easier to understand. [2] |

> **WJEC**
> An application question with 4 marks could only be used in Wales.

# Dealing with complex questions

Some questions seem to be very long and wordy. Don't panic. Break the questions down into parts to understand what the examiner wants you to do. In each question, look for the following:

- The command. This is usually (but not always) the first word in the question – look again at the examples in **Figure 3**.
- Instructions to use a figure. This will be a photo, map, graph or some text in the exam paper that contains useful clues. You **must** refer to the evidence provided.
- Whether you need to write about more than one thing. For example, a question could be about economic **and** social reasons for migration. Sometimes students do the first part (economic, in this example) and forget to do the second (social) so they don't finish the question.

Use a highlighter pen or underline key words in each complex question in your exam paper. It's a great way to check that you understand the question before you start to answer it. An example is shown in **Figure 4**.

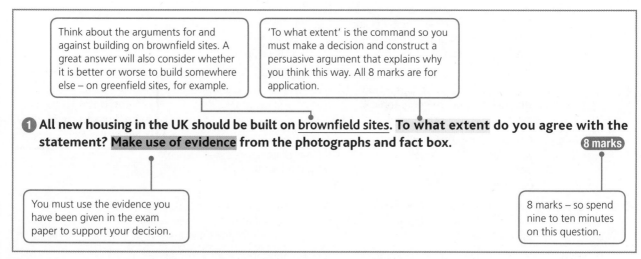

Figure 4 shows how to break down a complex question into its parts:

Think about the arguments for and against building on brownfield sites. A great answer will also consider whether it is better or worse to build somewhere else – on greenfield sites, for example.

'To what extent' is the command so you must make a decision and construct a persuasive argument that explains why you think this way. All 8 marks are for application.

**1** All new housing in the UK should be built on brownfield <u>sites</u>. **To what extent** do you agree with the statement? **Make use of evidence** from the photographs and fact box.   **8 marks**

You must use the evidence you have been given in the exam paper to support your decision.

8 marks – so spend nine to ten minutes on this question.

**Figure 4 How to break down a complex question into its parts**

HUG the question! Sometimes candidates seem to write everything they know about a subject, without actually answering the question! To avoid this, **HUG** the question:

**Highlight** the command word.

**Underline** other important instructions.

**Glance** back at the question to make sure you are actually answering it!

# Writing with accuracy

One 8-mark question in each paper has extra marks (3 in Wales, 4 in England) for writing with accuracy in your spelling, punctuation and grammar (SPaG).

It's worth taking a little extra time and care over this question. You should:
- write in full sentences (not bullet points)
- use paragraphs to give your answer a structure
- break up longer sentences using commas
- make sure you start each sentence with a capital letter
- check your spelling and correct it if necessary.

# Knowledge questions

Questions that test your **knowledge** of geography have 1, 2, 3 or 4 marks. Answers are point-marked. This means the examiner gives 1 mark for each correct point that you make.

## Fill-in-the-gaps questions

**Figure 5** shows an example of one of these questions. It's worth 4 marks so, if you need it, you can spend up to four minutes studying the diagram. There is no need to rush. You can revise for these kinds of questions by:

■ matching geographical terms to their definitions
■ practising drawing and labelling diagrams of geographical processes.

**Figure 5 An example of a 'fill-in-the-gaps' knowledge question**

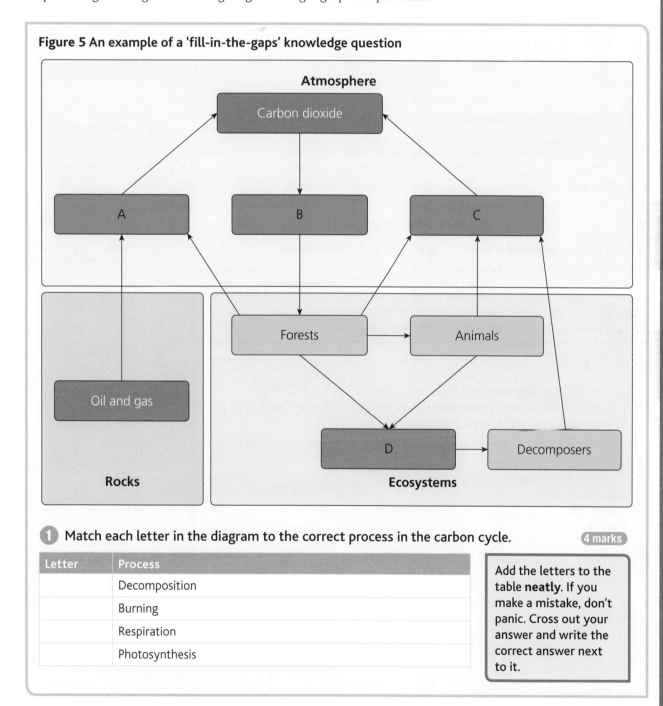

1  **Match each letter in the diagram to the correct process in the carbon cycle.** `4 marks`

| Letter | Process |
|--------|---------|
|        | Decomposition |
|        | Burning |
|        | Respiration |
|        | Photosynthesis |

Add the letters to the table **neatly**. If you make a mistake, don't panic. Cross out your answer and write the correct answer next to it.

# 'Describe' questions

Questions that have the command 'Describe' are usually worth 2, 3 or 4 marks. These questions also test your **knowledge** of geography. Answering a 'Describe' question is simple – make the same number of points as there are marks. If there are 3 marks, make three points. **Figure 6** shows an example of a question and mark scheme.

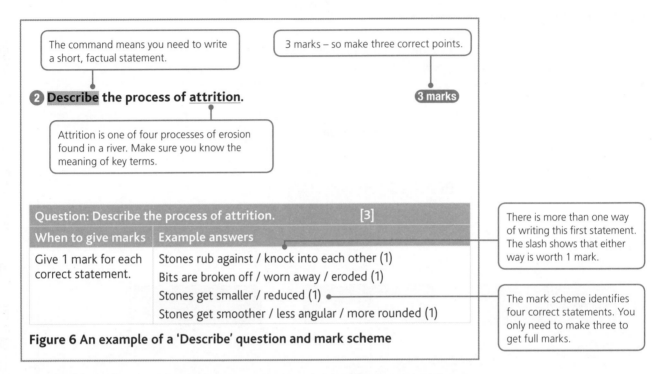

The command means you need to write a short, factual statement.

3 marks – so make three correct points.

**2 Describe** the process of <u>attrition</u>.

**3 marks**

Attrition is one of four processes of erosion found in a river. Make sure you know the meaning of key terms.

| Question: Describe the process of attrition. | [3] |
|---|---|
| When to give marks | Example answers |
| Give 1 mark for each correct statement. | Stones rub against / knock into each other (1) |
| | Bits are broken off / worn away / eroded (1) |
| | Stones get smaller / reduced (1) |
| | Stones get smoother / less angular / more rounded (1) |

There is more than one way of writing this first statement. The slash shows that either way is worth 1 mark.

The mark scheme identifies four correct statements. You only need to make three to get full marks.

**Figure 6** An example of a 'Describe' question and mark scheme

Keep your answers short like the example answers in the mark scheme. Don't give reasons – you won't get any extra marks by trying to explain.

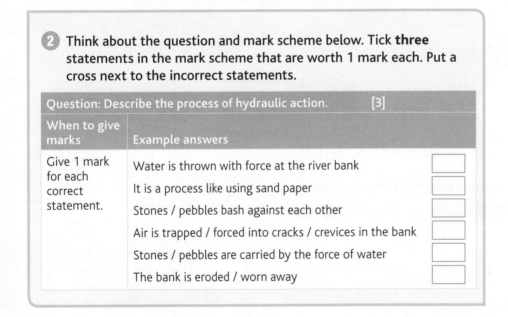

**2** Think about the question and mark scheme below. Tick **three** statements in the mark scheme that are worth 1 mark each. Put a cross next to the incorrect statements.

| Question: Describe the process of hydraulic action. | [3] | |
|---|---|---|
| When to give marks | Example answers | |
| Give 1 mark for each correct statement. | Water is thrown with force at the river bank | |
| | It is a process like using sand paper | |
| | Stones / pebbles bash against each other | |
| | Air is trapped / forced into cracks / crevices in the bank | |
| | Stones / pebbles are carried by the force of water | |
| | The bank is eroded / worn away | |

# Understanding questions

Questions that test your **understanding** of geography usually have 2, 4, or 6 marks. Only a few commands are used:

Give **one** reason    Give **two** reasons    Explain why

## Give one/two reasons

The simplest understanding questions use the command 'Give one reason'. This question is usually worth 2 marks. To answer it you need to make a point (P) and then explain (E) it. The examiner will give 1 mark for a correct point plus 1 mark for its explanation. Look at the example of a question and the student responses in **Figure 7**.

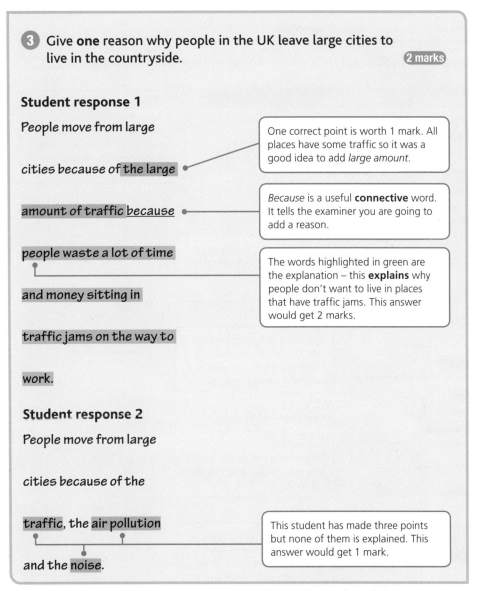

**3** Give **one** reason why people in the UK leave large cities to live in the countryside.    (2 marks)

**Student response 1**

People move from large

cities because of the large

> One correct point is worth 1 mark. All places have some traffic so it was a good idea to add *large amount*.

amount of traffic because

> *Because* is a useful **connective** word. It tells the examiner you are going to add a reason.

people waste a lot of time

and money sitting in

> The words highlighted in green are the explanation – this **explains** why people don't want to live in places that have traffic jams. This answer would get 2 marks.

traffic jams on the way to

work.

**Student response 2**

People move from large

cities because of the

traffic, the air pollution

> This student has made three points but none of them is explained. This answer would get 1 mark.

and the noise.

**Figure 7** An example of a 'Give one reason' question and student responses

**WJEC**
In Wales, the maximum mark for an understanding question is 8 marks.

**Eduqas**
In England, the maximum mark for an understanding question is 6 marks.

# 'Explain why' questions

The 'Explain why' questions test your understanding – whether you can give detailed reasons. To answer them you need to make connections, for example, by linking a cause with its effects. 'Explain why' questions worth over 4 marks are marked in a banded mark scheme rather than by individual points.

You can start by making a simple point (P) and then add an explanation (E) just like with a 2-mark question. However, you need to develop higher quality answers that show a **depth** of understanding when you answer a question worth 6 marks in England or 8 in Wales. The use of **connectives** will help you to link ideas together and write better answers that go into more depth. The easiest connective to use is 'so'. To illustrate this, let's think about how to answer the question shown in **Figure 8**.

**WJEC**
In Wales, 'Explain why' questions could be worth 4, 6 or 8 marks.

**Eduqas**
In England, 'Explain why' questions are worth 4 or 6 marks.

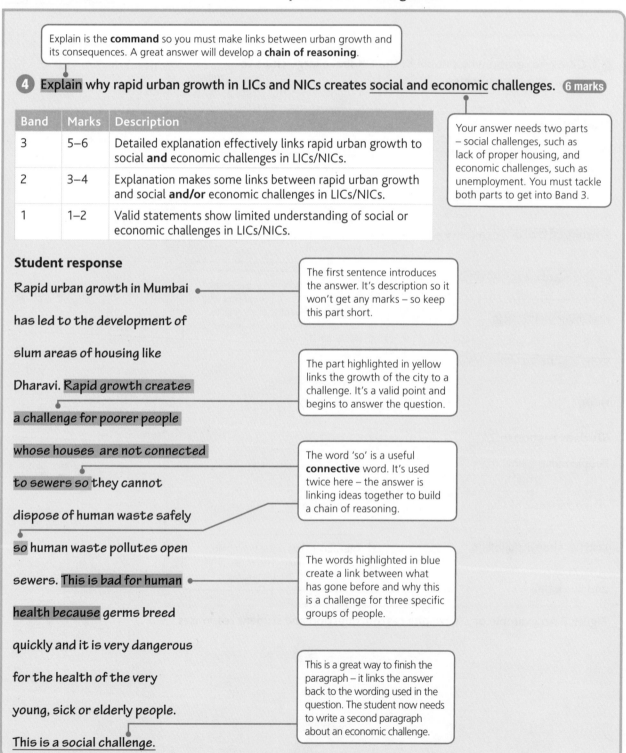

Explain is the **command** so you must make links between urban growth and its consequences. A great answer will develop a **chain of reasoning**.

**4** **Explain** why rapid urban growth in LICs and NICs creates social and economic challenges. **6 marks**

| Band | Marks | Description |
|---|---|---|
| 3 | 5–6 | Detailed explanation effectively links rapid urban growth to social **and** economic challenges in LICs/NICs. |
| 2 | 3–4 | Explanation makes some links between rapid urban growth and social **and/or** economic challenges in LICs/NICs. |
| 1 | 1–2 | Valid statements show limited understanding of social or economic challenges in LICs/NICs. |

Your answer needs two parts – social challenges, such as lack of proper housing, and economic challenges, such as unemployment. You must tackle both parts to get into Band 3.

**Student response**

Rapid urban growth in Mumbai has led to the development of slum areas of housing like Dharavi. Rapid growth creates a challenge for poorer people whose houses are not connected to sewers so they cannot dispose of human waste safely so human waste pollutes open sewers. This is bad for human health because germs breed quickly and it is very dangerous for the health of the very young, sick or elderly people. This is a social challenge.

The first sentence introduces the answer. It's description so it won't get any marks – so keep this part short.

The part highlighted in yellow links the growth of the city to a challenge. It's a valid point and begins to answer the question.

The word 'so' is a useful **connective** word. It's used twice here – the answer is linking ideas together to build a chain of reasoning.

The words highlighted in blue create a link between what has gone before and why this is a challenge for three specific groups of people.

This is a great way to finish the paragraph – it links the answer back to the wording used in the question. The student now needs to write a second paragraph about an economic challenge.

**Figure 8** An example of an 'Explain why...' question, mark scheme and students response

## Developing chains of reasoning

To develop an extended answer you should make a point and then ask yourself the question, 'So what?' This technique forces you to explain the consequences of the simple statement, creating longer sentences that show you fully understand something. Study the first row of **Figure 9**. Notice how the connective 'so' has been used to link ideas together. A great answer develops these chains of reasoning.

> Asking yourself 'So what?' is a simple trick that will help you to improve your answers. It forces you to add explanation. Do it every time you make a simple point until you have created a chain of reasoning.

⑤ Explain why rapid urban growth in LICs and NICs creates social and economic challenges.
**6 marks**

In **Figure 9**, a student has made a list of simple points that could help to answer the question. Use your understanding of the issues facing cities in LICs and NICs to complete each of the explanations in **Figure 9**.

Developing a chain of reasoning ➡

| Point | Explanation | Further explanation |
|---|---|---|
| 1 Houses are often badly built without proper foundations… | …so houses built on a slope are at risk of collapse during the rainy season… | …so people are at risk of losing their homes and all their possessions. |
| 2 The houses are made of recycled materials like corrugated tin… | …so they leak water during the monsoon season… | |
| 3 There is a lot of rubbish left lying around… | …so rats could be attracted… | |
| 4 There are too few jobs in the formal sector… | | |
| 5 There aren't enough teachers in the secondary school… | | |
| 6 There isn't enough street lighting… | | |

**Figure 9 Use the connective 'so' to turn simple statements into a chain of reasoning**

# Skills questions

Questions that test your use of geographical **skills** have 1, 2, 3 or 4 marks. Most answers are point-marked. This means the examiner gives 1 mark for each correct point that you make.

Skills questions are about:
- reading maps – including OS maps
- using graphs
- doing simple calculations

## Using graphs

Graphs are used to present geographical data. Graph questions may ask you to:
- read a value from the graph
- complete the graph by adding a data point
- describe the shape (pattern or trend) of the graph.

Exam papers may contain a variety of different graphs including **bar charts**, **line graphs**, **climate graphs** and **scatter graphs**.

> To describe a climate graph you need to give an **overview** of the whole year. You can calculate the temperature **range** by finding the difference in temperature between the hottest month and coldest month. You can describe **patterns** of rainfall by describing dry seasons (when rainfall is below average) and wet seasons (when rainfall is above average).

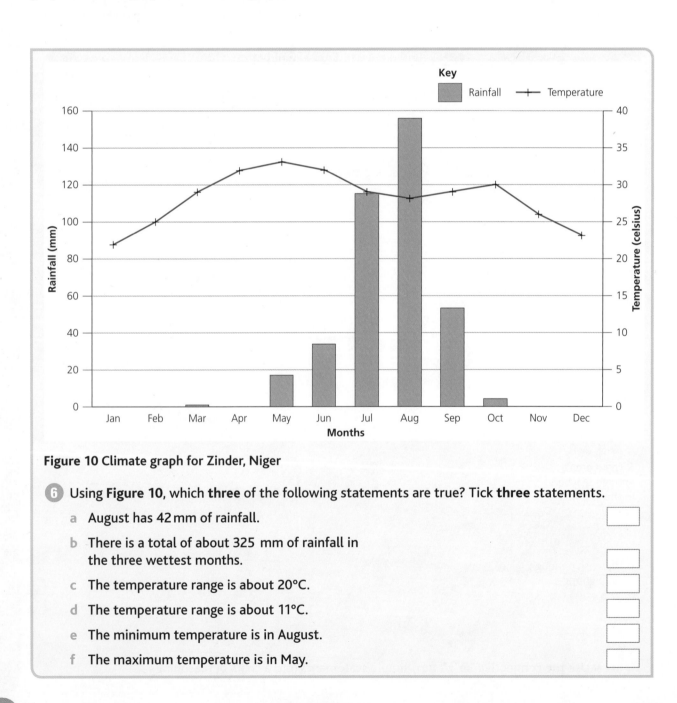

**Figure 10 Climate graph for Zinder, Niger**

6 Using **Figure 10**, which **three** of the following statements are true? Tick **three** statements.

a August has 42 mm of rainfall.

b There is a total of about 325 mm of rainfall in the three wettest months.

c The temperature range is about 20°C.

d The temperature range is about 11°C.

e The minimum temperature is in August.

f The maximum temperature is in May.

# Selecting and adapting graphs

As well as being able to **use** graphs (by reading, completing or describing them) you also need to be able to:

■ select an appropriate style of graph to use

■ comment on how a graph could be adapted or improved.

## Selecting graphs

The style of graph you draw depends on the type of data that is being presented. Certain rules apply. For example, **discrete** (or **categorical** data) should be presented using a bar chart and never with a line graph. If the examiner gives you a table of data, you need to be able to select an appropriate graph and be able to justify (give reasons) for your decision. **Figure 11** gives some advice on graph selection.

**Figure 11 Which graph goes with which data?**

| Type of data | Example | Suitable graphs to select | Justification |
|---|---|---|---|
| Discrete | Data that can be counted and put into categories, e.g. types of vehicle in a traffic survey or different locations for a pedestrian count. | Bar charts (vertical or horizontal) or divided bars | Each category is different so needs to be presented using a separate bar. |
| Continuous | Data that can be measured, such as wind speed, water depth, plant height or distance along a transect. | Line graphs or cross-sections | Values are numerical and vary continuously so a line is the best way to present the continuous changes. |
| Percentage | Discrete data where we are interested in the proportion of the population that has a particular characteristic, e.g. the percentage of people in a ward who are retired. | Pie charts or pictograms | Pie charts are best because they are a visual presentation of proportion rather than absolute size. This means they are useful for making comparisons, e.g. between wards of different sizes. |
| Pairs of data | Data where one variable varies in relation to another variable, such as wind speed and altitude. | Scatter graphs | Scatter graphs present a visual scatter of points that show a possible correlation between two sets of data. |

## Adapting graphs

The examiner may ask you how a graph could be adapted, changed or improved. In these questions, check the graph carefully to make sure that the data has been presented using a suitable style of graph. You must also check that the graph has all of the information that it should have. You should check that:

■ it has a title and a key

■ each axis is labelled and has units of measurement

■ that each axis starts at zero.

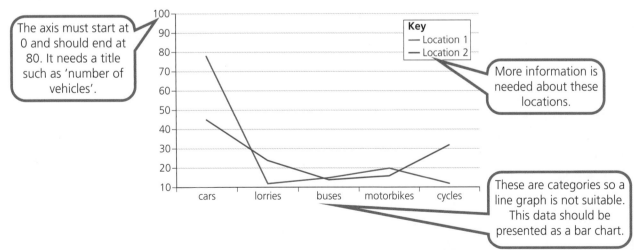

**Figure 12 How could this graph be adapted/improved?**

# Reading OS maps

Ordnance Survey (OS) map extracts are used to show the shape of the UK's landscape. Two scales are often used in exam papers:

- 1:50,000 scale where each 2 cm grid square represents 1 km² on the ground.
- 1:25,000 scale where each 4 cm grid square represents 1 km² on the ground.

In order to answer questions which use an OS map you need to be able to use:

- four-figure and six-figure **grid references** to locate places on the map
- the **scale line** to calculate distance and **compass arrow** to work out directions
- contour lines to understand the **relief** (the shape of the landscape).

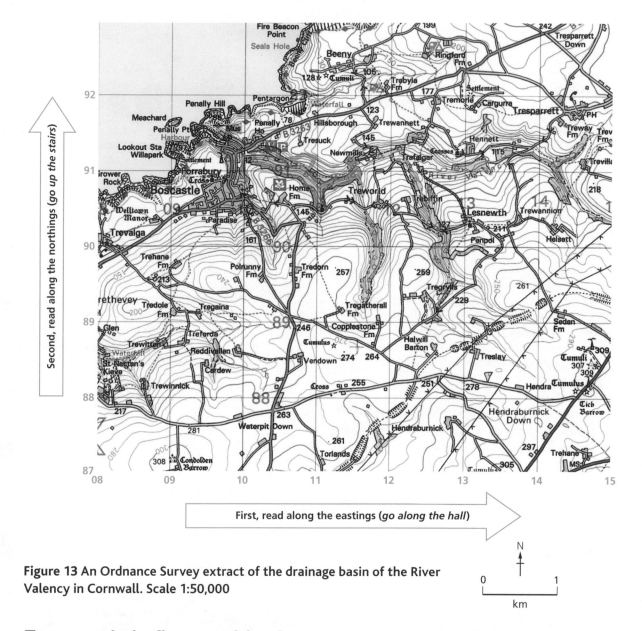

**Figure 13 An Ordnance Survey extract of the drainage basin of the River Valency in Cornwall. Scale 1:50,000**

## Four- and six-figure grid references

Four-figure grid references identify a complete grid square. Tregatherall Fm (a farm) is to the south-east of Boscastle in the southern half of the map. It has a four-figure reference of 1189.

Six-figure grid references identify a specific location on the map. To turn a four-figure into a six-figure reference you need to do the following:

**1** Find the third digit by estimating how many tenths of a kilometre the location is past the last easting. Tregatherall Farm is three-tenths past easting 11, so the first three digits are 113.

**2** Find the sixth digit by estimating how many tenths of a kilometre the location is past the last northing. Tregatherall Farm is two-tenths past northing 89, so the last three digits are 892.

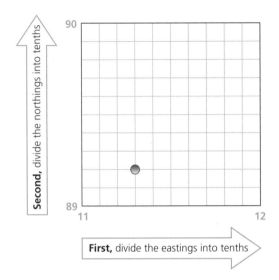

**Figure 14 Finding the third and sixth number in a six-figure grid reference: the location of Tregatherall Farm**

Always 'go along the hall before you go up the stairs' to get your grid reference numbers in the correct order.

---

**7** Match the following grid references to features in **Figure 13**.

**1390**       **0991**       **137895**       **115911**       **104897**

| Grid reference | Feature |
|---|---|
| | Source of a stream close to Poirunny Farm |
| | Spot height of 261 m |
| | The village of Lesnewth |
| | Confluence of River Valency and a north flowing tributary |
| | Mouth of the River Valency |

**8** Which is the best description of grid square 1189? Tick one answer.

a   An upland area with gentle slopes and several streams.

b   A steep-sided river valley. The river flows in a northerly direction. The valley is wooded.

c   A flat, lowland area with a river valley. The valley is wooded.

**9** Describe the relief of the area covered by this map.

.........................................................................................

.........................................................................................

.........................................................................................

.........................................................................................

Use the contour lines to imagine the shape of the land. The closer the contour lines, the steeper the gradient of the slope.

# Reading other maps

Exam papers contain a variety of different styles of maps. Some, like **Figure 15**, show where features are located.

Features (or data values) on a map may show a pattern. If so, the question may ask you to **describe the distribution**. In your answer, you need to describe the pattern carefully. The following words are helpful:

- **clustered**: points on a map are concentrated into small groups
- **linear**: features on a map are spread out along lines
- **random**: features are at irregular distances from each other. There is no clear pattern
- **regular**: features are spaced out evenly across the map.

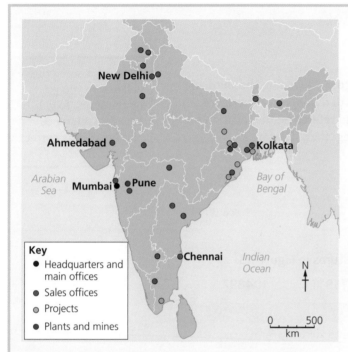

**Figure 15 Tata steel factories and offices in India**

10 Using **Figure 15**, which **one** of the following statements best describes the location of Kolkata?

a  Near the coast of the Bay of Bengal.

b  North of the Indian Ocean and east of New Delhi.

c  1600 km ENE of Mumbai.

d  1600 km WSW of Mumbai

11 Using **Figure 15**, describe the location of Chennai.

.................................................................

.................................................................

> Never describe somewhere as 'near to'. It's too vague. You should **always** use a compass direction and distance from an important point on the map.

12 Using **Figure 15**, describe the distribution of Tata sales offices.

.................................................................

.................................................................

> Use words like **clustered**, **random** and **regular** to describe distribution. This is an opportunity to show the examiner your use of geographical vocabulary.

# Doing simple calculations

Some questions test your ability to process geographical data by asking you to do simple calculations. The data will be presented in the exam paper as a table, graph or map. The question will use a command word such as '**calculate**'. They are simple questions which are usually worth 2 or 3 marks. You are allowed to use a calculator to find the answer.

An example of a numeracy question is one that asks you to calculate a **percentage**. A percentage is a way of expressing a part of a whole. Look at **Figure 16**. There are 46 megacities (cities with over 10 million people) in the world. The whole number in this case is 46 and this is 100 per cent. If 23 of these cities were all in the same continent, that would be 50 per cent of the whole. To calculate the percentage of megacities that are in China, follow these steps (**Figure 17** shows how you could show your working):

- **Step 1:** Divide the number of megacities in China (15) by the total number of megacities (46).
- **Step 2:** Multiply the answer from Step 1 by 100.
- **Step 3:** You can round the number up or down to the nearest whole number or to 1 decimal place (dp). If the number ends .49 or less, round it down. If it ends .50 or more, round it up. So 15.37 would round down to 15.4 per cent (1 dp) or 15 per cent (to the nearest whole number).

> **Do** show how you got the answer if the question says 'Show your working'. In a 2-mark question, you may get 1 mark by showing your working, even if your answer is wrong.

**Figure 16 Number of world megacities (cities with a population greater than 10 million) in each continent**

| Continent | | Number of megacities |
|---|---|---|
| Africa | | 3 |
| Asia | China | 15 |
| | India | 5 |
| | Rest of Asia | 12 |
| Europe | | 3 |
| North America | | 3 |
| South America | | 5 |
| **Total number** | | **46** |

**Figure 17 How to calculate a percentage**

Number of megacities in China is 15 $\quad \dfrac{15}{46} = 0.32608$

$$0.32608 \times 100 = 32.608$$

$$= 32.6 \text{ (to 1 d.p)}$$

13 Using **Figure 16**, calculate the percentage of all megacities in each of the following places. Show your answer to 1 dp.

| a Rest of Asia | b India | c Europe |
|---|---|---|
| | | |

# Application questions

Application questions assess your ability to **infer**, **analyse**, **evaluate** or use evidence to **make a decision**. Most application questions give you some evidence to study. This could be a photo, Fact Box or map which you need to read alongside the question. Part of the test is to see how well you can read this evidence (a skill known as inference) and use it to support your answer. There are at least as many marks for application as any other type of question – so it's important to understand how these questions work. Only a few commands are used:

| Suggest | Evaluate | Analyse | To what extent |
|---------|----------|---------|----------------|

**WJEC**
In Wales, application questions can be worth 4, 6 or 8 marks.

**Eduqas**
In England, application questions in Papers 1 and 2 are always worth 8 marks. There is one at the end of each theme.

## Questions that use photos

Photos are often used in geography exams. The photos are not there to make the exam paper look pretty! They provide important geographical evidence so study each one carefully. Question setters often use photos to test whether you can 'read' the geography that can be seen in the photo.

### Point, evidence, inference

Look at **Figure 18**. It shows damage caused by a river flood to shops in Cockermouth in Cumbria. You almost certainly won't have seen this photo before, but you should have learned about the impacts of river floods and you may have seen similar photos. You should be able to interpret the evidence you can see in this photo by using what you learned in your geography lessons. The label next to **Figure 18** has been added by the author of this book – you won't find helpful labels like this around the photos in the exam paper! You need to ask yourself some simple questions about what you can **infer** from the evidence in the photo:

> What could…?   How might…?   What ought…?

Study the photo below.

> Use evidence in the photo, e.g. the boarded-up shops or the rubbish in the skip. What can you infer? Think about what this evidence could tell you about the impacts on the economy. The impacts could be good or bad.

**Figure 18** How to deal with a question that uses a photo

**14** Suggest how this river flood affected the economy. Use evidence from the photograph.

> 'Suggest' means you need to propose a possible answer based on what you can see. Think about how this flood will have economic effects on businesses, shoppers, the emergency services or local builders. Ask yourself questions, such as 'How might local businesses be affected?'

**WJEC**
In Wales, a **suggest** question would usually be worth 4 marks.

**Eduqas**
In England, this question would be worth 8 marks. There could be some extra evidence to look at, such as a map of the drainage basin.

| Make a **point** | Provide some **evidence** | Make an **inference** |

| Flood water has damaged businesses in this street. | Floor boards have been ripped up and thrown in the skip. | The flood has created work for local builders who have been employed to make repairs. |

Use the **Point, Evidence, Inference** technique to add to the answer to Question 14. Think about a negative impact on the economy.

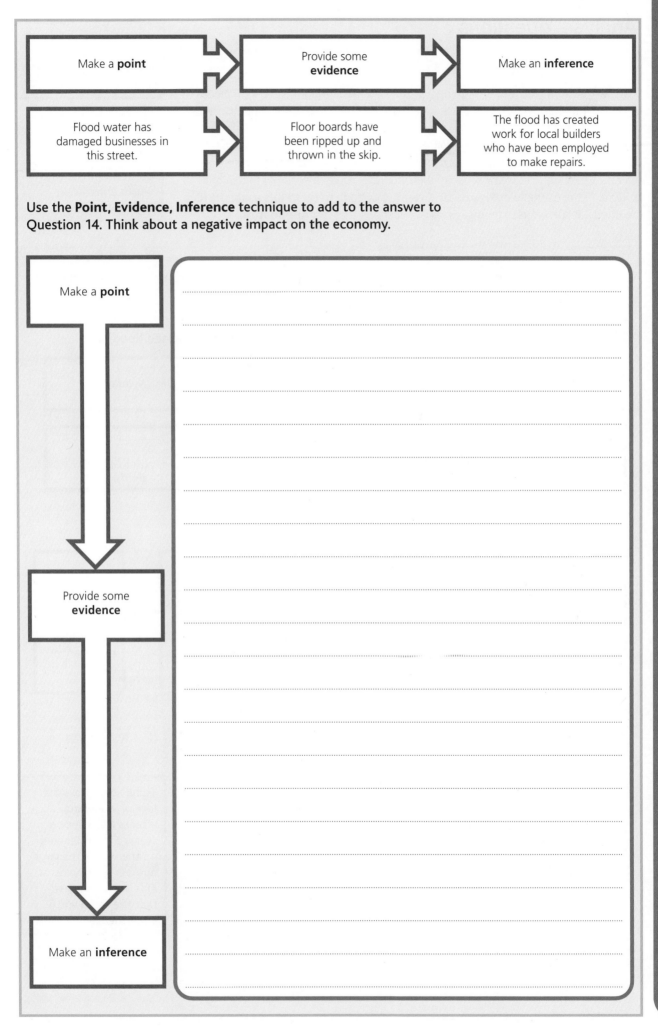

Make a **point**

Provide some **evidence**

Make an **inference**

# 'Analyse' questions

To **analyse** means to break down the evidence so that you can see the overall pattern or reason for something.

Analyse questions are a bit like doing a jigsaw. You need to look at several pieces of evidence and then link them all back together with some reasoning. The examiner will be looking for explanations, so a good way to tackle an analyse question is with the 'So what?' technique. A great answer will link together some **chains of reasoning** (page 11). A really great answer will also use the Point, Evidence, Inference technique to make sense of the evidence. You can see how this might be done in the example in **Figure 19**. Overall, this book refers to this as the **PEIC** technique. It can be used to answer most questions that test **application**.

**WJEC**
In Wales, analyse questions could be worth 6 or 8 marks.

**Eduqas**
In England, analyse questions are always worth 8 marks on Papers 1 and 2.

15 Analyse the factors that increase the risk of flooding in the drainage basin of the River Valency. Use evidence from the OS map on page 14.

**Step 1:** Use **PEI** to identify and infer how one factor on the map may have caused flooding.

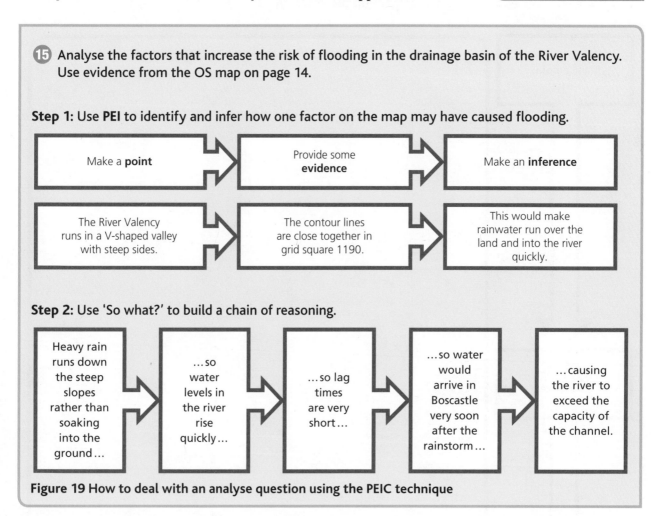

**Step 2:** Use 'So what?' to build a chain of reasoning.

**Figure 19 How to deal with an analyse question using the PEIC technique**

16 Analyse how **two** of the following factors may increase the risk of flooding of the River Valency. Use the PEIC technique each time. Use evidence from Figure 13, page 14.

  a  The fact that there is hardly any woodland in the drainage basin.

  b  The River Valency has lots of small tributaries.

  c  Upland areas (over 250 m) are very close to Boscastle.

Think about how the lack of woodland affects interception storage. Another clue on the map is the large number of streams. This would suggest that the rocks are impermeable.

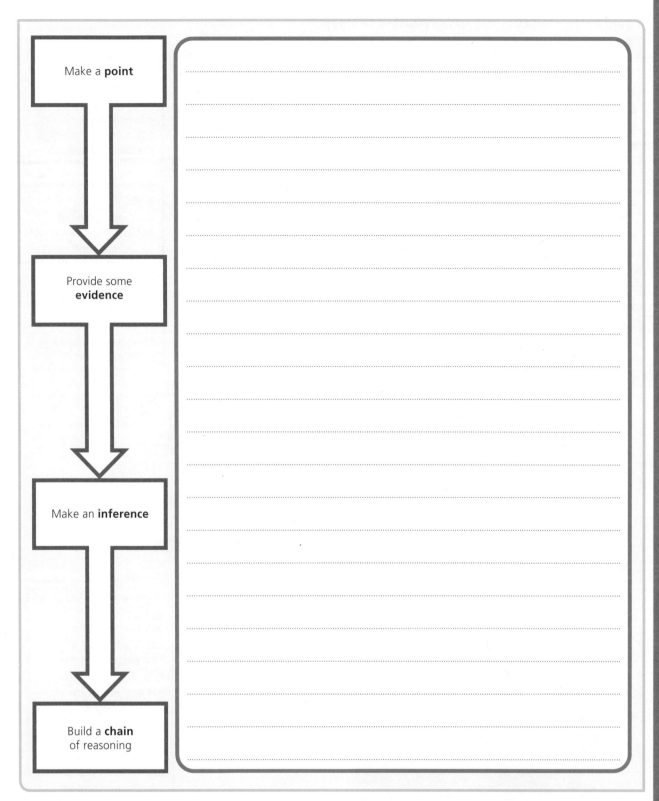

## 'Evaluate' questions

To **evaluate** means to identify strengths and weaknesses or advantages and disadvantages.

You can evaluate lots of geography! You can practise by making lists of the advantages and disadvantages of management strategies you have studied in class. For example, you could evaluate the strategies used to manage a honeypot site or you could evaluate a coastal management scheme. Use a table like **Figure 20**. Think about the impacts of the strategy on people (**social** impacts), the **environment** and the **economy** – the **SEE** technique.

**WJEC**
In Wales, evaluate questions could be worth 4, 6 or 8 marks.

**Eduqas**
In England, evaluate questions are always worth 8 marks in Papers 1 and 2.

**Figure 20 Use the SEE technique to evaluate a strategy**

| | Advantages | Disadvantages |
|---|---|---|
| SOCIAL (for people and communities) | | |
| ENVIRONMENTAL (water use, ecosystems, air pollution) | | |
| ECONOMIC (jobs and businesses) | | |

A good evaluation will look at advantages **and** disadvantages. A great answer will come to a conclusion – do the advantages outweigh the disadvantages?

## 'To what extent' questions

'To what extent' means you need to use **evidence** to help you **make a decision**. This type of question will be accompanied by at least one map, graph, photo or Fact Box – or a combination of these types of evidence. This evidence will give you some information about a place – an example is shown in **Figure 21**. You may have never studied this place in class but that doesn't matter. You will have studied somewhere else that faces a similar issue or problem. You need to apply your understanding (from your learning in class) to help you make the decision in this new situation.

**WJEC**
In Wales, 'To what extent' questions could be worth 6 or 8 marks.

**Eduqas**
In England, 'To what extent' questions are always worth 8 marks in Papers 1 and 2.

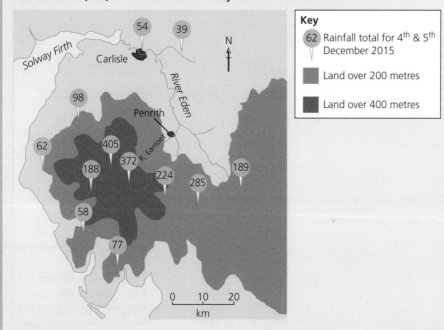

### Fact Box

Very heavy rainfall in Cumbria between 4th December and 6th December 2015 caused major floods in Carlisle.
Average rainfall in Penrith is 215 mm for the whole of December.
Flood defences costing £38 million were built in Carlisle between 2005 and 2010.
More than 2,000 homes were damaged in Carlisle during the 2015 flood.

**Rainfall totals (mm) in Cumbria over two days in December 2015**

Key
62 Rainfall total for 4th & 5th December 2015
Land over 200 metres
Land over 400 metres

'To what extent' questions sometimes include a Fact Box. Don't just copy out sentences from the Fact Box – you won't get any marks for copying. You need to make use of this evidence. The best way to do this is to ask yourself the 'So what?' question so that you demonstrate that you can infer the consequences of this evidence.

**Figure 21 An example of a 'To what extent' question. The map is based on data from the Met Office website.**

17 'It is impossible to prevent all floods in Carlisle.' To what extent do you agree with this statement? Use evidence from the Fact Box and map.

## Structuring your answer

A good answer to this question will have three parts to it.

- **An argument**. This paragraph will use evidence that supports the view. Use the PEIC technique. In this example you might argue that it is impossible to protect people from floods after extreme weather events like the one of December 2015. Use evidence to support your argument.
- **A counter-argument**. Use PEIC again to consider the opposing point of view. In this example, you might argue that better management of rivers in the upper drainage basin would store water for longer and prevent any flooding downstream in Carlisle.
- **A decision**. Your final paragraph should weigh up the evidence, and reach a decision about whether it is possible to prevent the effects of *every* flood. You might come to a straightforward *yes* or *no* – a black-and-white decision. Alternatively, it's okay to argue for something in between. If so, use the 'washing line' technique, shown in **Figure 23**, to help you word your decision.

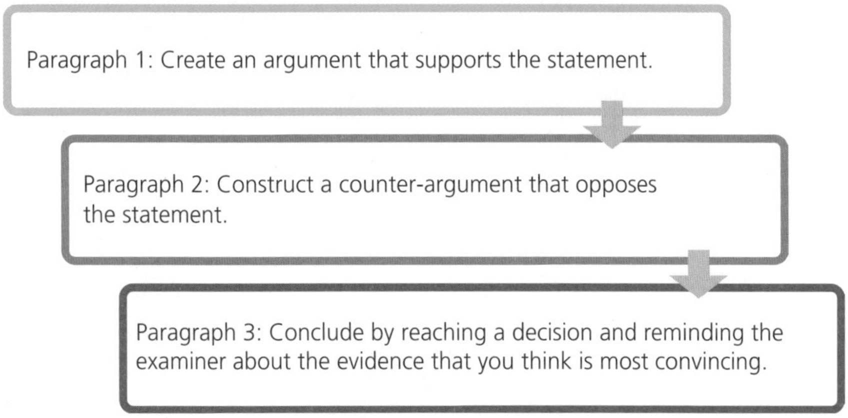

Paragraph 1: Create an argument that supports the statement.

Paragraph 2: Construct a counter-argument that opposes the statement.

Paragraph 3: Conclude by reaching a decision and reminding the examiner about the evidence that you think is most convincing.

**Figure 22 The structure for an answer to a question that asks you to make a decision**

## What is the examiner looking for in your answer?

A good answer will include the following:

- An answer to the question – in this case whether you think it is possible to prevent floods or not.
- Direct reference to **evidence** in the photo and Fact Box. In this example, the evidence clearly shows that the upper part of the catchment area for the river that flows through Carlisle had an exceptional amount of rainfall.
- Clear **connections** to your own learning when you are dealing with the issue – in this case, flooding and the difficulty of trying to prevent river floods.
- A short **conclusion**.

A great answer will also do the following:

- Consider both sides of the argument. This is what the examiner calls **balance**.
- Include **chains of reasoning** that show you have detailed understanding of why flooding occurs and why it is difficult and expensive to prevent.

### Signposting your answer

You can use signposting to help structure your answer in a way that the examiner will find helpful and clear. Signposting is a technique that tells the reader what is coming next – like a signpost tells you where you are going. Here are a few useful signposts you can use.

| To signpost an argument: | To signpost a counter-argument: | To signpost your conclusion: |
|---|---|---|
| On the one hand…<br>One view would be… | On the other hand…<br>In comparison…<br>Another possibility is… | Overall, I think…<br>My conclusion is… |

# Dealing with 'To what extent?'

You **must** state whether you agree with the statement or not. You may fully agree or disagree with the statement. It's also possible that you only partially agree with it. Either way, it doesn't really matter because the examiner isn't looking for a particular answer. It's the way that you use the evidence to support your decision that is important. **Figure 23** gives you some helpful phrases to use in your answer.

> Even if you fully agree or completely disagree, you should **always** present both sides of an argument and then make a decision.

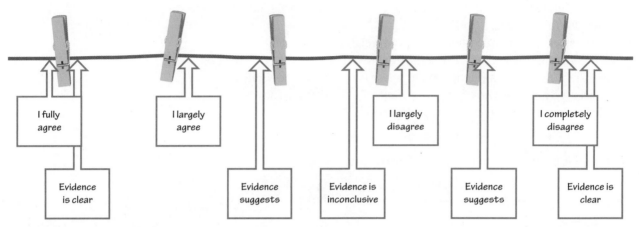

**Figure 23 Use a 'washing line' to help you state your decision**

## Writing a conclusion

You should finish your answer to a 'To what extent' question with a short conclusion. **Figure 24** gives you some advice.

**Figure 24 Dos and don'ts of writing a conclusion**

| Do | Don't |
|---|---|
| Do remind the examiner that you have looked at both sides of the argument. | Don't sit on the fence! If you have been asked to make a judgement, you should make it. |
| Do repeat what you think is the most persuasive or significant piece of evidence that supports your decision. | Don't state that it is difficult to make a decision if the earlier part of your essay is packed full of strong evidence. |
| Do use key words or phrases from the question in your conclusion. | Don't forget to glance back at the question before you start the conclusion. If you have wandered off task then now is your chance to save your essay and answer the question! |

**18** 'It is impossible to prevent all floods in Carlisle.' To what extent do you agree with this statement? Use evidence from the Fact Box and map on page 22.

**Argument**

**Counter-argument**

**Conclusion**

# Chapter 2 Preparing for Paper 1

## Theme 1 Landscapes and physical processes

Theme 1 is a core theme so you **must** revise it. It is examined in Paper 1, Question 1.

### Distinctive landscapes

Each **landscape** in Wales and the UK has its own special character. Factors that make landscapes distinctive are **geology** (rocks), **relief** (shape of the landscape), **land use**, **vegetation** and **culture** (the history of people who have lived there).

It's difficult to see **geology** – unless you can see a cliff – but it's an important part of the landscape. It is usually the hardest rocks that form our mountains. Softer rocks, like clay, often form valleys.

**Figure 1 Aerial photo of Monmouth, Wales**

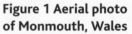

People make a **cultural** impact on the landscape. For example, by using different building materials and distinct styles of architecture in different parts of the UK.

**Figure 2 Aerial photo of Wylfa Nuclear Power Station, Anglesey, North Wales**

1. Study **Figure 1** and **Figure 2**. Make notes about what makes these landscapes distinctive using the table below.

| | Figure 1 | Figure 2 |
|---|---|---|
| Relief | | |
| Land use; e.g. type of farms and size of fields | | |
| Culture, e.g. towns and buildings | | |
| Vegetation, e.g. woodland | | |

2. Use your notes to **compare** these landscapes. Make three different points.

..........................................................................................................

..........................................................................................................

..........................................................................................................

..........................................................................................................

..........................................................................................................

..........................................................................................................

..........................................................................................................

..........................................................................................................

..........................................................................................................

..........................................................................................................

..........................................................................................................

> If you are asked to **compare** you must make some direct links between the two things – in this case two landscapes. It isn't good enough to describe one landscape and then the other one. Use words like *whereas* or *similarly* to connect sentences. Also use words like *smaller*, *steeper* or *higher* to make comparisons.

# Managing landscapes

Some landscapes have a lot of visitors. These landscapes are called **honeypot sites**. They need to be carefully managed. One way to do this is by giving them status as **Areas of Outstanding Natural Beauty (AONBs)**.

**3** Which key geography term is being described here? Underline **one** term.

**Definition:** A place that attracts so many visitors that it may be damaged.

**honeypot site**          **brownfield site**          **greenfield site**

**Figure 3** Visitors to this upland landscape have eroded a footpath

**4** Study **Figure 3** and the phrases below.

| | |
|---|---|
| A  Plant roots hold the soil together. | F  Soil is eroded by rainwater. |
| B  Trampling kills plants. | G  The path gets wider. |
| C  People walk to the summit. | H  Wet soil sticks to the boots of walkers. |
| D  Leaves intercept rainwater. | I  Undamaged plants. |
| E  People avoid walking on uneven stones. | |

a   Put each of the letters A–I in the correct boxes on **Figure 3**. You can put more than one letter in each box.

b   Use the photo and phrases to make a **chain of reasoning** that answers the **explain** question below:

Explain why footpaths are eroded at honeypot sites.

> Use the 'So what?' technique to help you make a chain of reasoning that uses as many of the phrases as possible. See page 10-11.

# River and coastal landforms

The processes of **erosion**, **transportation** and **deposition** cause landforms to change over time. Make sure you can:

- use the correct technical terms
- explain how each process causes the landform to change.

**5** There are many geographical terms to learn for this theme. These terms describe river processes.

  a  Circle **one** word that is an erosion process.

  b  Underline **one** word that is a transportation process.

  c  Match **three** of the terms to the correct definition.

abrasion      hydraulic action      lateral erosion      vertical erosion      traction

| Process | Description |
| --- | --- |
| | When pebbles are rolled along the bed of a stream by the force of the water flowing in the river. |
| | The sweeping motion of the water in a river which widens each meander sideways. |
| | When stones are thrown against the banks of the river causing some of the bank to be worn away. |

**Figure 4 A large meander on the River Severn**

**6** Label the following features on **Figure 4**.

  A  **slip-off slope**

  B  **river cliff**

  C  **floodplain**

  D  **river channel**

Study **Figure 4** and the phrases below.

water flows slowly      the channel is deep      less energy      the channel is shallow

water flows quickly      unable to transport sediment      lots of energy

more friction between the water and the river bed      the channel is efficient

**7** Use the photo and phrases to make a **chain of reasoning** that answers the **explain** questions:

  a  Explain why rivers erode on the outside bend of a meander.

  ..........................................................................................................................

  ..........................................................................................................................

  ..........................................................................................................................

  b  Explain why rivers deposit sediment on the inside bend of a meander.

  ..........................................................................................................................

  ..........................................................................................................................

  ..........................................................................................................................

**8** The following terms describe coastal processes.

a Circle **one** word that is an erosion process.

b Underline **one** word that is a transportation process.

c Match **three** of the terms to the correct definition.

Use the 'So what?' technique to help you make a chain of reasoning that uses as many of the phrases as possible.

**longshore drift   solution   attrition   hydraulic action   slumping**

| Process | Description |
| --- | --- |
|  | A sudden movement of soil and unconsolidated (loose) rocks down a cliff face. |
|  | When pebbles bang into each other making them smaller and more rounded. |
|  | When waves forcefully throw water against a cliff face causing some of the cliff to be worn away. |

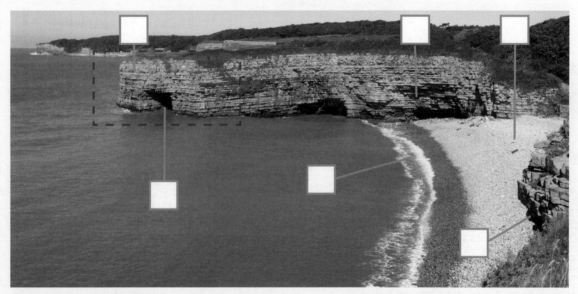

**Figure 5 A coastal landscape in Glamorgan**

**9** Put each of the letters A–F in the correct box on **Figure 5**.

A **beach**       B **bedding planes**       C **headland**       D **cave**

E **overhang**       F **swash/backwash**

**10** Suggest how the landform inside the red dotted box may change over time.

..................................................................................................................

..................................................................................................................

..................................................................................................................

..................................................................................................................

..................................................................................................................

..................................................................................................................

..................................................................................................................

..................................................................................................................

Make sure you use geographical terms when you answer this type of question. Process terms like *abrasion*, *hydraulic action* and *rock fall* or *recession* could be useful. You should also use terms for features such as *bedding planes*, *arches* or *wave-cut platforms*.

# Landform rates of change

All landforms change shape. For example, **erosion** causes cliffs to collapse so the coastline **recedes**. Landforms usually change very slowly, but sometimes change can be rapid. Factors that affect the rate of landform change are **geology** (rocks), **weather and climate** and **human activity**.

## Geology

Unconsolidated (loose) rocks like clay erode more easily than solid rocks like limestone or granite which are harder (more **resistant**). If a river flows from a harder rock onto a softer rock it will begin to erode more rapidly, forming a waterfall and gorge. The rate of change of the coastline also depends on geology. Resistant rocks make headlands that retreat slowly. Softer rocks make bays that retreat more rapidly.

> It's better to describe rocks as harder rocks and softer rocks rather than hard and soft. Even better, describe rocks as resistant to erosion and less resistant to erosion.

⑪ Draw a diagram to show what happens when rivers flow from harder to softer rocks. Imagine **Figure 6** is the first drawing in a cartoon. Show how you expect the River Mellte has changed. Label your diagram with at least **three** key terms to describe important features and processes.

The river flows across a fault line from harder rock to softer rock

Harder rock

Softer rock

Fault

> If the exam question is about a landform, you can include a sketch as part of your answer. The best sketches are carefully labelled to explain how geographical processes change the landform over time.

**Figure 6 What the Afon (River) Mellte used to look like**

# Weather and climate

Storms caused by low pressure create strong winds and larger waves. These conditions may cause cliffs to collapse.

**12** Match **six** key terms to the correct definition below. **Two** terms are not needed.

| fetch | prevailing wind | storm surge | longshore drift |
|---|---|---|---|
| cliff recession | discharge | low pressure | high pressure |

| Term | Definition |
|---|---|
| | A weather system that causes strong winds. |
| | Erosion which causes the coastline to retreat. |
| | The direction from which the wind often blows. |
| | A measure of the distance that has been travelled by the wind across a sea or ocean. |
| | Extra-large waves and high sea levels caused by low air pressure. |
| | A measure of the amount and speed of water flowing in a river. |

Study **Figure 7** and the phrases below.

**discharge is greater**
**the channel is deeper**
**more energy**
**larger pebbles can be transported**
**river cliffs collapse**
**more sediment falls into the river**
**more force**
**more traction**

**Figure 7** Evidence of slumping as soft soils of the floodplain fall into the river

**13** Use the photo and phrases to make a **chain of reasoning** that answers the **explain** questions below:

> Use the 'So what?' technique to help you make a chain of reasoning that uses as many of the phrases as possible.

   **a** Explain why a river is able to carry more pebbles after a period of heavy rainfall.

.................................................................................................................................

.................................................................................................................................

.................................................................................................................................

.................................................................................................................................

   **b** Explain why more erosion occurs in a river after heavy rainfall.

.................................................................................................................................

.................................................................................................................................

.................................................................................................................................

.................................................................................................................................

.................................................................................................................................

# Human activity

Sometimes river or coastal management, which is meant to reduce flooding or erosion, causes problems elsewhere. These are called **unintended consequences**.

**14** Link each point to one effect and one unintended consequence using an arrow to make three simple chains of reasoning.

| Point | Effect | Unintended consequence |
|---|---|---|
| When a river is straightened | …water moves with less friction and has more energy | …so further along the coast the beaches are thinner and offer less protection to the cliffs. |
| When a river has smooth concrete embankments | …sediment is prevented from moving along the coast | …so the river channel further downstream is more likely to suffer erosion. |
| When groynes are built on the coast | …water takes less time to travel downstream | …so places downstream can be affected by flooding. |

**Figure 8** A large rock groyne at Mappleton on the Holderness coast of Yorkshire

**15** Add annotations to each box on **Figure 8**. Use your annotations to:

a  describe the coastal processes on either side of the groyne

b  explain why the groyne has had unintended consequences.

# Flooding

As water moves through a **drainage basin** its speed of flow depends on a number of factors such as **climate**, **vegetation** and **geology**. In some situations, the flow of water causes river **flooding**. People manage flooding through **hard engineering**, **soft engineering** and **land use zoning** of the floodplain.

16 Match **four** key terms to the correct definition below. **Two** terms are not needed.

discharge    infiltration    interception    throughflow    evaporation    impermeable

| Term | Definition |
|---|---|
|  | Materials such as clay and tarmac that prevent water passing through. |
|  | Flows of water from the surface of the soil into the ground. |
|  | Movement of water down a slope through the soil. |
|  | Stores of water on the leaves of plants and canopies of trees. |

17 Each description (A–F) below describes a different drainage basin. Decide whether each drainage basin would have a hydrograph like X or Y. Copy each letter into the correct column of the table below.

A   Impermeable rocks like clay

B   Lots of urban areas

C   Lots of woodland

D   Porous rocks like sandstone

E   Hardly any urban areas

F   Hardly any woodland

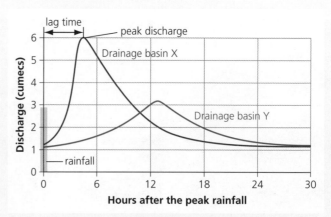

**Figure 9 Hydrographs for two contrasting drainage basins**

| Drainage basin X | Drainage basin Y |
|---|---|
|  |  |

# Example of a flood

Use the table below to summarise **one** example of a flood you have studied. Use bullet points to list key facts.

| River: | Place that flooded: |
|---|---|
| Cause of the flood, e.g. heavy rainfall or snow melt |  |
| Effects of the flood, e.g. on people, business or transport |  |

**Figure 10. Flood management strategies used in the city of York**

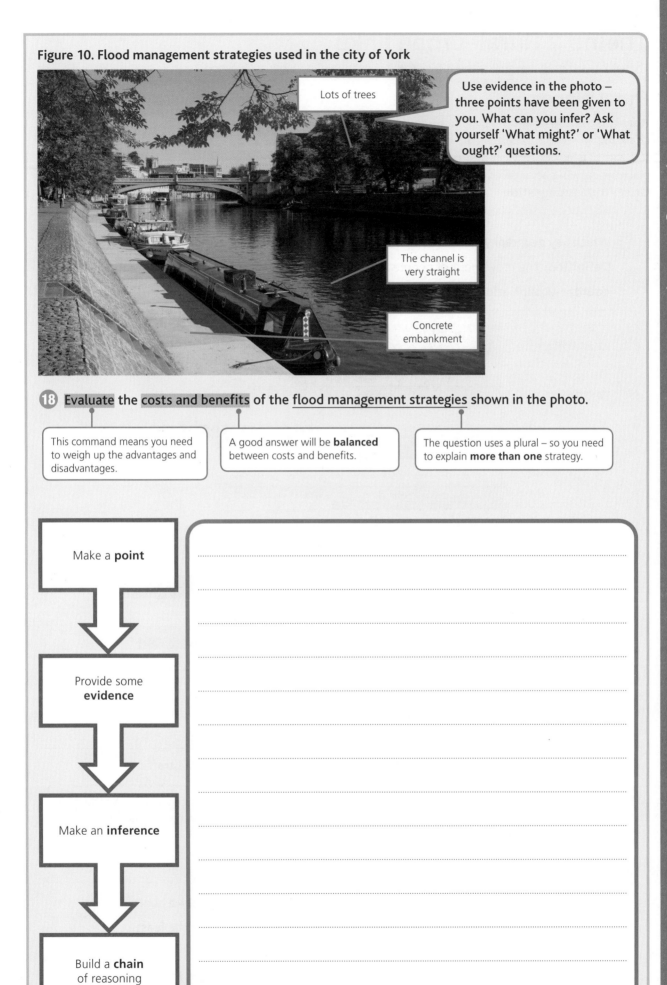

Lots of trees

Use evidence in the photo – three points have been given to you. What can you infer? Ask yourself 'What might?' or 'What ought?' questions.

The channel is very straight

Concrete embankment

**18** Evaluate the costs and benefits of the flood management strategies shown in the photo.

This command means you need to weigh up the advantages and disadvantages.

A good answer will be **balanced** between costs and benefits.

The question uses a plural – so you need to explain **more than one** strategy.

Make a **point**

Provide some **evidence**

Make an **inference**

Build a **chain** of reasoning

# Theme 2 Rural–urban links

Theme 2 is a core theme so you **must** revise it. It is examined in Paper 1, Question 2.

## Rural places

**Rural** places (towns and villages in the countryside) in Wales and England are changing. In some rural places the population is falling. People move away because of a lack of services and **rural poverty**. The population of other rural places is growing because of **counter-urbanisation** and **commuting**. This causes traffic **congestion**.

1. Which key geography term is being described here? <u>Underline</u> one term.

   **Definition:** The movement of people and some jobs from cities to smaller towns and villages.

   counter-urbanisation       commuting       congestion

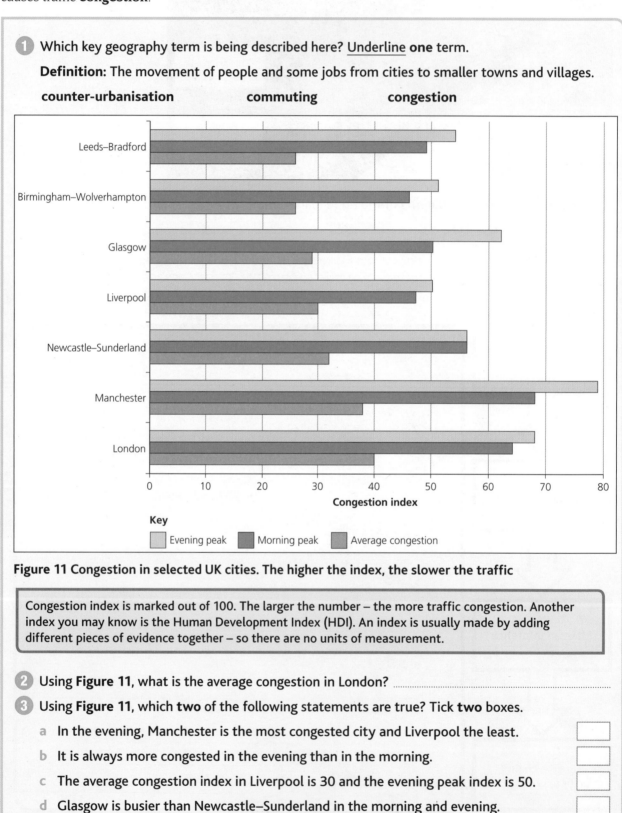

**Key**
Evening peak   Morning peak   Average congestion

**Figure 11 Congestion in selected UK cities. The higher the index, the slower the traffic**

Congestion index is marked out of 100. The larger the number – the more traffic congestion. Another index you may know is the Human Development Index (HDI). An index is usually made by adding different pieces of evidence together – so there are no units of measurement.

2. Using **Figure 11**, what is the average congestion in London? ..................................

3. Using **Figure 11**, which **two** of the following statements are true? Tick **two** boxes.

   a   In the evening, Manchester is the most congested city and Liverpool the least. ☐

   b   It is always more congested in the evening than in the morning. ☐

   c   The average congestion index in Liverpool is 30 and the evening peak index is 50. ☐

   d   Glasgow is busier than Newcastle–Sunderland in the morning and evening. ☐

④ Match the factors (A–E) that cause counter-urbanisation to the explanations in the table below. One has been done for you.

A  Changes in technology       C  Social factors              E  Economic factors

B  Changes to employment       D  Environmental factors

| Factor | Point | | Explanation |
|---|---|---|---|
| | Fibre optic cable is being laid in towns in rural areas of the UK | | ...so people are concerned about their health especially if they have asthma. |
| | Work places allow staff to work flexible hours | | ...so people can work from home and communicate with work colleagues. |
| | Some cities have higher crime rates than rural areas | | ...so people can come into work later to avoid rush hour traffic. |
| | Some inner cities have very high levels of air pollution | | ...so people can travel to jobs in the city centre without having to drive or park. |
| | Cars are safer, more comfortable and more economical than before | | ...so people move to the countryside where they feel safer. |
| E | House prices (and rent) in London are very high | | ...so people are prepared to drive further to work than ever before. |
| | Many workers commute long distances using rail links | | ...so people move to cheaper homes and commute back into London. |

> Demand for houses in the countryside is growing and rural house prices are high. In many places rural homes are dearer than urban homes. London is an exception.

## Urban change in the UK

Lack of **housing** is a major **challenge** in many UK towns and cities. Should new housing be built on **greenfield** or **brownfield sites**?

⑤ What are the advantages and disadvantages of building new homes on brownfield or greenfield sites? Sort these statements into the table below by writing the letters A–H in the correct places.

A  People want to live close to the city centre.

B  People think that the countryside should be kept for farming and wildlife.

C  Land is polluted so it is expensive to clean it up before building starts.

D  Derelict land is often in an inner urban area. Building new housing here makes the inner city overcrowded.

E  It is cheaper to build on an open site with no dereliction.

F  The site already has gas, electricity and sewers nearby.

G  People want to have a home on the edge of smaller towns because it is quieter.

H  Building in the countryside adds to traffic congestion because people have to commute.

| | Advantages | Disadvantages |
|---|---|---|
| Greenfield sites | | |
| Brownfield sites | | |

# Population change in the UK

In the UK we are living longer so the population is **ageing**. The population is changing in other ways too as people are moving to find work. These **migrants** are often young adults. Many migrants move from towns in **rural** areas to cities where they start a family. This means new housing needs to be built, especially in London and the south-east region of the UK, which are places where the population is growing fastest.

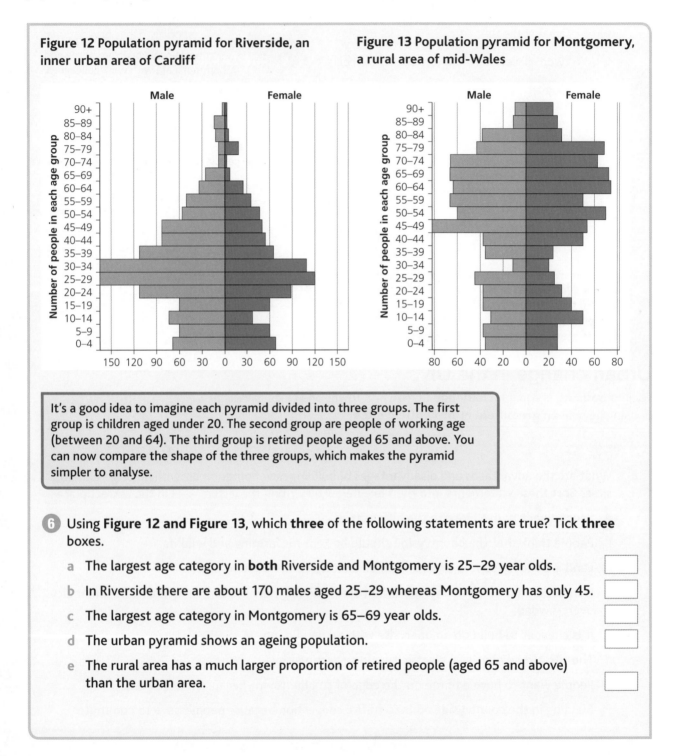

**Figure 12** Population pyramid for Riverside, an inner urban area of Cardiff

**Figure 13** Population pyramid for Montgomery, a rural area of mid-Wales

It's a good idea to imagine each pyramid divided into three groups. The first group is children aged under 20. The second group are people of working age (between 20 and 64). The third group is retired people aged 65 and above. You can now compare the shape of the three groups, which makes the pyramid simpler to analyse.

**6** Using **Figure 12 and Figure 13**, which **three** of the following statements are true? Tick **three** boxes.

a   The largest age category in **both** Riverside and Montgomery is 25–29 year olds.  ☐

b   In Riverside there are about 170 males aged 25–29 whereas Montgomery has only 45.  ☐

c   The largest age category in Montgomery is 65–69 year olds.  ☐

d   The urban pyramid shows an ageing population.  ☐

e   The rural area has a much larger proportion of retired people (aged 65 and above) than the urban area.  ☐

Study **Figures 12 and 13** and the phrases below.

retirement    peaceful rural environment    crowded cities    younger adults

better access to healthcare facilities    better paid jobs in cities

universities in cities    lack of jobs    lack of affordable housing

closure of village schools    online banking    fewer rural bus services

**7** Use **Figures 12 and 13** and the phrases to make a **chain of reasoning** that answers the **explain** questions below:

a  Explain why many rural areas of the UK have an ageing population.

..................................................................................................................

..................................................................................................................

..................................................................................................................

..................................................................................................................

..................................................................................................................

..................................................................................................................

b  Explain why more family homes need to be built in many urban areas.

..................................................................................................................

..................................................................................................................

..................................................................................................................

..................................................................................................................

..................................................................................................................

..................................................................................................................

c  Explain why access to services is more difficult in rural areas than in urban areas.

..................................................................................................................

..................................................................................................................

..................................................................................................................

..................................................................................................................

..................................................................................................................

..................................................................................................................

..................................................................................................................

# Retail change

The way we shop is changing. More people are shopping online or in out-of-town shopping centres rather than in town centres. This has **costs** and **benefits** for UK towns and cities.

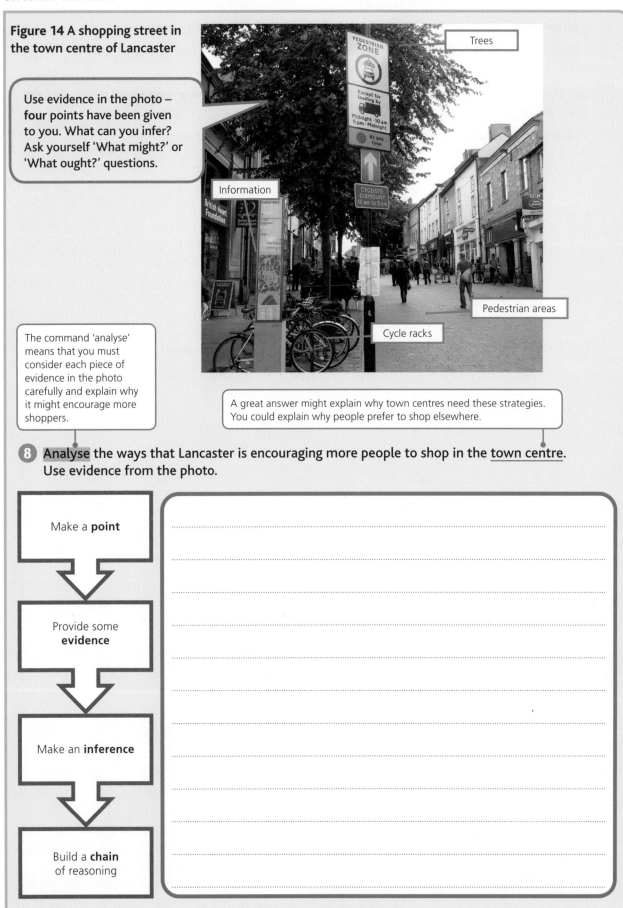

**Figure 14** A shopping street in the town centre of Lancaster

Trees

Information

Pedestrian areas

Cycle racks

Use evidence in the photo – four points have been given to you. What can you infer? Ask yourself 'What might?' or 'What ought?' questions.

The command 'analyse' means that you must consider each piece of evidence in the photo carefully and explain why it might encourage more shoppers.

A great answer might explain why town centres need these strategies. You could explain why people prefer to shop elsewhere.

8 **Analyse** the ways that Lancaster is encouraging more people to shop in the <u>town centre</u>. Use evidence from the photo.

Make a **point**

Provide some **evidence**

Make an **inference**

Build a **chain** of reasoning

# Global patterns of urbanisation

This theme uses some important abbreviations:

- **Low income countries (LICs)** are very poor countries. Most LICs are in sub-Saharan Africa, such as Sierra Leone, Uganda and Tanzania. Towns and cities (**urban** places) are growing fastest in LICs.
- **Newly industrialised countries (NICs)** are developing countries which have a lot of manufacturing industry. Many NICs are in Asia (such as India and China) or South America (e.g. Brazil). The world's largest cities are found in NICs, for example, Mumbai, Shanghai and São Paulo. Any city with over 10 million people is a **megacity**.
- **High income countries (HICs)** are very wealthy countries such as the UK, Japan, Australia and the USA. There are some **megacities** in HICs, such as London, Tokyo and New York.

9 Which key geography term is being described here? <u>Underline</u> **one** term.

**Definition:** Cities that are well connected to other places around the world.

**megacities**      **capital cities**      **global cities**

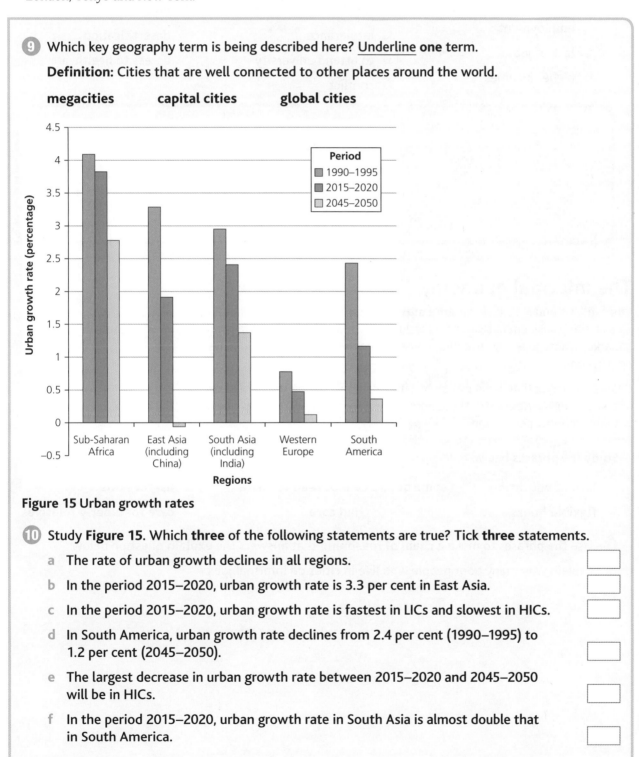

**Figure 15** Urban growth rates

10 Study **Figure 15**. Which **three** of the following statements are true? Tick **three** statements.

a The rate of urban growth declines in all regions.

b In the period 2015–2020, urban growth rate is 3.3 per cent in East Asia.

c In the period 2015–2020, urban growth rate is fastest in LICs and slowest in HICs.

d In South America, urban growth rate declines from 2.4 per cent (1990–1995) to 1.2 per cent (2045–2050).

e The largest decrease in urban growth rate between 2015–2020 and 2045–2050 will be in HICs.

f In the period 2015–2020, urban growth rate in South Asia is almost double that in South America.

## Push–pull theory

Urban areas in LICs and some NICs continue to grow rapidly due to:

- **migration** – people moving to urban areas from the countryside (**rural** places)
- **natural increase** – more births than deaths.

Migration happens because of a mixture of:

- **push factors** – reasons you want to leave your home
- **pull factors** – reasons that attract you to move to somewhere else.

> Never give 'income' or 'jobs' as pull factors. People are pulled by the idea they might get a **higher** income or because there will be **more** jobs available.

**11** Sort the following list into push factors and pull factors by placing each letter in the correct box.

| | | |
|---|---|---|
| A shortage of food | E flooding | J to join family |
| B seasonal work/under-employment | F too far to nearest school | K drought |
| | G intolerance | L desertification |
| C lack of jobs | H to attend university | M access to healthcare |
| D higher incomes | I conflict | |

| Push factors | Pull factors |
|---|---|
| | |

## The informal economy

Cities in LICs and NICs have an **informal economy**. People working in informal jobs do things like street cleaning, recycling, rickshaw riding or selling goods on a market. These jobs have irregular hours, no sickness benefits, no holidays and no pensions.

Cities grow faster than housing can be built, resulting in the growth of **slums** and **squatter settlements**. These homes are also informal. They have been built without planning permission.

Study the phrases below.

| lack of education | some adults cannot read or write | useful skills |
|---|---|---|
| flexible hours | child care | care for elderly relatives |

**12** Use the phrases to make a **chain of reasoning** that answers the **explain** question below:

Explain why many poor people who live in cities do informal jobs.

...................................................................................................................................

...................................................................................................................................

...................................................................................................................................

...................................................................................................................................

...................................................................................................................................

...................................................................................................................................

# Global cities

**Global cities** are connected by **trade** through a major port or airport or through **finance** and business because they have offices of **multi-national companies (MNCs)**. Global cities are also connected through the movement of people (**migrants**). The growth of global cities creates opportunities for people:

- **Social opportunities** – better access to schools, clean water and healthcare.
- **Economic opportunities** – large cities have better global links, for example, international airports, so they have more industry.

## Examples of global cities

You will have studied two global cities. One in an LIC or NIC and the other in an HIC. Use the table below to summarise these examples. Use bullet points to list key facts.

| | LIC/NIC city | HIC city |
|---|---|---|
| Name | | |
| Patterns, e.g. areas of wealth and poverty | | |
| Why the city has grown, e.g. migration | | |
| Challenges, e.g. poverty or lack of housing | | |
| How the city is connected to the rest of the world | | |

If you are asked to describe your global city, be specific. For example, to describe how a city is connected to the rest of the world, name the international airport or give the name of an MNC that is located there.

**13** a  Explain why global cities create opportunities by linking the **Points** on the left to the **Explanations** on the right. One example has been done for you.

b  Use a highlighter to show how each explanation has been extended further.

| Points | Explanations |
|---|---|
| Global cities have international airports and ports ... | ... so there are a range of jobs in manufacturing with further opportunities for training and promotion for employees. |
| Universities are located in global cities ... | ... so ordinary people have access to safe drinking water, which means there are fewer deaths at a young age. |
| Multinational companies build factories and offices in global cities ... | ... so there are opportunities for firms that sell goods overseas to grow, which creates more jobs for local people. |
| Global cities in LICs and some NICs continue to grow ... | ... so there are opportunities for people to get degrees or vocational qualifications, which means they are more employable. |
| It is easier to provide water for an urban population than for people living in remote rural places ... | ... so there is the opportunity to get healthcare quickly, which might prevent more serious health problems from developing. |
| You are never far away from a health clinic or hospital in a large city ... | ... so there are always jobs available on construction sites requiring all sorts of skilled workers, from labourers to civil engineers. |

The 'So what?' technique has been used to explain each point (and highlighted in yellow). Notice how the sentence has been extended again to add further explanation (highlighted in green).

## Fact Box

It is estimated that 6.5 million people live in Mumbai's slum housing.

About 50 per cent of Mumbai's slums are informal – people have no right to live there.

At least 50,000 people in Mumbai have no home at all.

Use evidence in the photo. Think about the challenges shown. For example, the way the houses are built or how people are living close to this dirty water. What can you infer? Ask yourself 'What might?' or 'What ought?' questions.

**Figure 16 Dharavi slum in Mumbai**

You need to make a decision and then justify it – which means giving reasons. A great answer will have some balance between why it is possible and why it may be impossible to provide enough affordable, good homes.

**14** **To what extent is it possible for Mumbai to provide affordable and good homes for everyone? Use evidence in the photo and Fact Box.**

You must refer directly to evidence and what can be inferred from it.

Make a **point**

Provide some **evidence**

Make an **inference**

Build a **chain** of reasoning

# Theme 3 Tectonic landscapes and hazards

Theme 3 is an optional theme. It is examined in Paper 1, Question 3.

Tick the box if you studied this theme.

## Tectonic processes

Tectonic hazards are caused by the movement of the plates that make up the Earth's crust. Plate movement creates large features on the Earth's crust such as **rift valleys** and **ocean trenches**.

**1** Match **five** key terms to the correct definition below. **Two** terms are not needed.

| seismometer | volcanic hotspot | constructive | ocean trench |
|---|---|---|---|
| convection current | rift valley | subduction | |

| Term | Definition |
|---|---|
| | A long and very deep depression in the sea floor close to a subduction zone. |
| | A process which destroys crust when it is pulled under another plate. |
| | Movement of magma beneath the crust which may cause plate movement. |
| | Plate margins where plates are moving away from one another. |
| | An area of magma below the crust which causes a lot of volcanic activity. |

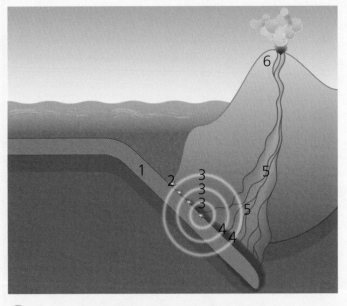

**Figure 17 The process of subduction occurs at destructive plate boundaries**

**2** Number each step in the following table to describe what happens as an oceanic plate is destroyed beneath a continental plate.

| Labels for Figure 17 | Number |
|---|---|
| The magma reaches the surface causing a volcanic eruption. | |
| There is friction between the oceanic and continental plates. | |
| Magma rises through the continental crust. | |
| The dense oceanic plate is pulled into the mantle. | |
| The heat and friction cause oceanic crust and ocean floor sediment to melt. | |
| The friction is overcome, causing an earthquake. | |

**3** Study the list of landforms below.

| | | | |
|---|---|---|---|
| shield volcano | cinder cone | ocean trench | stratovolcano |
| lava tube | caldera | rift valley | geyser |

a   Highlight two landforms that might be found at a destructive plate boundary.

b   Underline two landforms that might be found at a constructive plate boundary.

c   Circle the two features that are the largest in scale.

d   Match **four** of the landforms to the descriptions below.

| Landform | Description |
|---|---|
| | A large structure created by the collapse of a magma chamber. |
| | A small volcanic feature formed by the eruption of ash and volcanic bombs. |
| | A complex of tunnels created by hot, flowing lava. |
| | A small volcanic feature formed by the eruption of steam. |

## Examples of tectonic hazards

You will have studied examples of **one** volcanic hazard and **one** earthquake hazard. Use the space below to record the main facts about these examples.

| | Volcanic hazard | Earthquake event |
|---|---|---|
| Where | | |
| When | | |
| Impacts, e.g. effects on health, infrastructure and the economy | | |

# Impacts of tectonic processes

Tectonic processes create **hazards** which can have negative impacts on people. For example, an earthquake can cause serious injuries and other **health** issues if roads, water pipes, electricity and airports (**infrastructure**) are damaged and supplies of food and water are cut off.

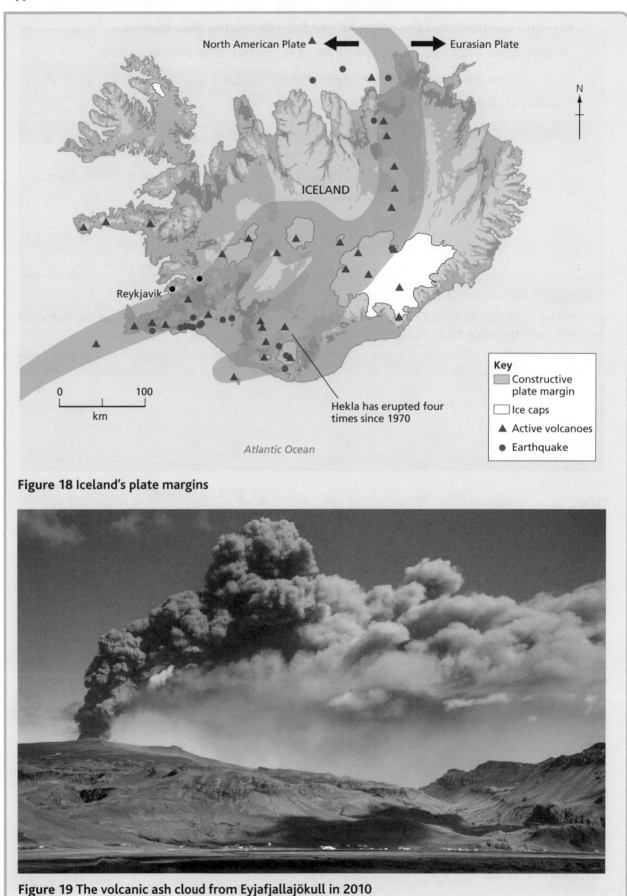

**Figure 18 Iceland's plate margins**

North American Plate ⬅ Eurasian Plate ➡

ICELAND

N

Reykjavik

Hekla has erupted four times since 1970

0    100
km

Atlantic Ocean

**Key**
- Constructive plate margin
- Ice caps
- ▲ Active volcanoes
- ● Earthquake

**Figure 19 The volcanic ash cloud from Eyjafjallajökull in 2010**

Suggest means you must use inference. Ask yourself 'What might?' or 'How could?' Think about impacts on health, the economy and infrastructure.

**4** **Suggest how tectonic activity could affect people living in Reykjavik and other parts of Europe. Use evidence in the map and photo.**

You **must** use evidence to support your answer. Use the PEIC technique. For example, you could work out the direction of the wind that would carry ash from Eyjafjallajökull over Reykjavik.

A great answer will deal with both. HUG the question so you don't miss this kind of instruction.

Make a **point**

Provide some **evidence**

Make an **inference**

Build a **chain** of reasoning

# Reducing the risk of tectonic hazards

We can reduce the risk of tectonic hazards through **monitoring** and **hazard mapping**. We can also reduce risk by improving **building technology** and **emergency planning**.

**5** Link the following phrases together with arrows to make a chain of reasoning.

| Point | Explanation | Further explanation |
|---|---|---|
| Better emergency planning | ... so people know which places are at risk of pyroclastic flows | ... so people can be evacuated to high ground in time |
| Monitor wave height | ... so fire crews and health workers work together effectively | ... so people are not crushed or trapped in collapsed buildings. |
| Better building technology | ... so places in danger of a tsunami can be given a warning | ... so people can be evacuated to safety before an eruption. |
| Create hazard maps | ... so buildings sway and flex during an earthquake | ... so people are given help quickly after an earthquake. |

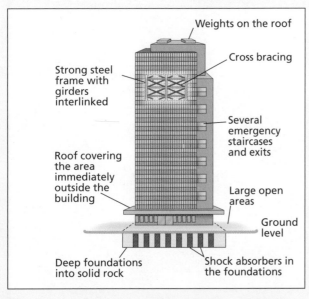

**Figure 20 An earthquake-resistant building**

**6** Choose **three** of the labels on **Figure 20**. For each of these, explain why the design feature would reduce the risk. Use the 'So what?' technique (page 11) to explain each point. One example has been done for you.

*A roof covers the area around the building* so *people are not injured by broken glass if it falls from windows.*

**Feature 1**

.......................................................................................................................................................

.......................................................................................................................................................

**Feature 2**

.......................................................................................................................................................

.......................................................................................................................................................

**Feature 3**

.......................................................................................................................................................

.......................................................................................................................................................

The command means you must make a decision – is it useful or not in reducing risk?

**7** **To what extent** can building technology be used to reduce risks from <u>earthquakes and volcanoes</u>? <u>Use evidence in **Figure 20**.</u>

You **must** use evidence to support your answer. Use the PEIC technique.

There must be two parts to your answer. You may try to argue that it is easier to protect people from earthquakes than from volcanic activity or vice versa.

Make a **point**

Provide some **evidence**

Make an **inference**

Build a **chain** of reasoning

# Factors that affect vulnerability

Volcanoes and earthquakes pose a risk to people. The amount of risk depends on **physical factors** such as the type of event (for example, whether it is a lava flow or pyroclastic flow) and the **magnitude** (or size) of the event. Risk also depends on **social and economic factors**. For example, people who are very poor are more **vulnerable** (less likely to cope) than other people.

**Figure 21** A shanty town of informal homes on the edge of Mexico City

concrete blocks

shallow foundations

poor planning control

weak mortar

steep slopes

building on soft soil

soil turns to liquid during earthquake

no building control

poor education

overcrowding

no emergency plan

8 Study **Figure 21** and the phrases above.

Use the photo and phrases to make a **chain of reasoning** that answers the **explain** questions below:

> Use the 'So what?' technique to help you make a chain of reasoning that uses as many of the phrases as possible.

a Explain why informal homes are vulnerable to earthquakes.

.............................................................................................................

.............................................................................................................

.............................................................................................................

.............................................................................................................

.............................................................................................................

.............................................................................................................

b Explain why the poorest people in low income countries (LICs) are most vulnerable to tectonic hazards.

.............................................................................................................

.............................................................................................................

.............................................................................................................

.............................................................................................................

.............................................................................................................

.............................................................................................................

# Theme 4 Coastal hazards and their management

Theme 4 is an optional theme. It is examined in Paper 1, Question 4.

Tick the box if you studied this theme.

## Factors that increase vulnerability

Coastlines can be at risk of **erosion** and **coastal flooding** during **storms** or due to **sea level rise**. If somewhere or someone is at risk we say they are **vulnerable**. Vulnerability depends on various factors:

- **Physical** factors such as the **weather**, **height above sea level** or **time of day**. For example, risk increases during stormy weather.
- **Human** factors such as **income**, **age**, **health** or **disability**. For example, people on low incomes who live in poor quality housing are more vulnerable to coastal flooding than wealthier people.

---

**1** How do human factors increase vulnerability to coastal hazards? Match each factor to an explanation.

| A | Time of day | C | Age | E | Population density |
|---|---|---|---|---|---|
| B | Health | D | Wealth | | |

| Factor | Explanation |
|---|---|
| | It is more difficult to organise a safe evacuation in a crowded place. |
| | Elderly people are more vulnerable because they depend on help to evacuate safely. |
| | Poor people may live in unsafe buildings (or caravans). |
| | Evacuation is more difficult and dangerous at night when it is dark. |
| | People who are unwell are at much greater risk after a disaster than a well person. |

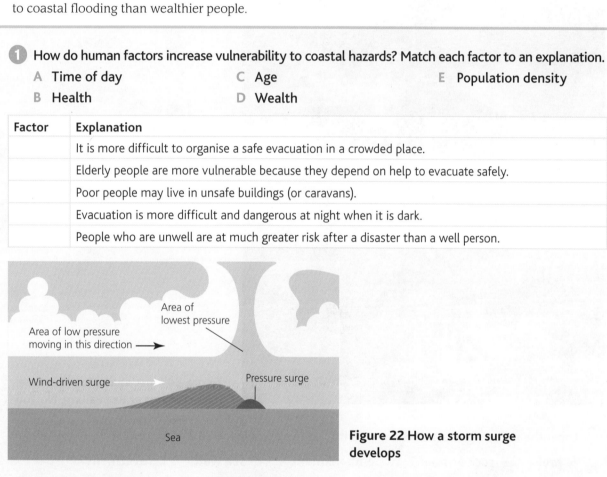

**Figure 22** How a storm surge develops

| very low air pressure | strong winds | less force from the atmosphere |
|---|---|---|
| sea levels rise | waves pile up | levels 3m higher than normal |

**2** Study **Figure 22** and the phrases above.

Use the photo and phrases to make a **chain of reasoning** that answers the **explain** question below:

Explain why coastal areas are vulnerable to flooding during storms.

........................................................................................................

........................................................................................................

........................................................................................................

........................................................................................................

# Coastal management

Coasts can be managed to try to prevent flooding and erosion. Two strategies can be used:

- **Hard engineering** strategies such as building **sea walls** or **groynes**.
- **Soft engineering** strategies such as **beach replenishment**.

If erosion can be prevented by building a sea wall, we are **holding the line**. Land is protected from erosion and flooding if the **benefits** are greater than the **costs**. Coastal management is expensive, so 'hold the line' is used if there are **economic** and **social** benefits to keeping the coastline as it is.

**Figure 23 Examples of coastal management at Sea Palling, Norfolk**

**Figure 24 More examples of coastal management at Sea Palling, Norfolk**

**3** Label the following coastal management strategies on **Figure 23** and **Figure 24**.

A **reef**    B **rock armour**    C **embankment**    D **sea wall**    E **revetment**

**4** Link each point to **one** explanation using an arrow to explain the coastal management at Sea Palling.

| Point | Explanation |
|---|---|
| The concrete sea wall is curved… | …so that they slow down in the shallow water which causes them to deposit sediment. |
| Waves wrap around the end of each reef… | …so that the strength of the backwash is reduced which prevents erosion of the beach. |
| Boulders absorb wave energy… | …so that the energy of waves at high tide is reflected back out to sea. |
| The sand dunes create a high embankment… | …so it creates a buffer and is able to absorb wave energy. |
| The beach is thicker and wider… | …so the lying land behind the beach is protected from coastal flooding. |

## Fact Box

- The Holderness coast erodes at an average rate of 2–3 metres per year.
- A policy of **hold the line** prevents erosion at Withernsea.
- Between 3 metres and 5 metres of coastline are eroded each year to the south of Withernsea.
- Withernsea has a population of 6,200.
- It has several primary schools and one secondary school.
- Withernsea has two holiday parks.

Never copy information from a Fact Box (examiners call this 'lifting') unless you can add some value to it. Add value by asking yourself 'So what?'

What can you infer?

How might the groynes affect coastal processes to the south of Withernsea?

What might it cost to build and maintain these coastal defences?

**Figure 25 The beach at Withernsea on the Holderness coast of Yorkshire**

**5** Evaluate the costs and benefits of holding the line at Withernsea. Use evidence in the photo and Fact Box.

To answer this question you must weigh up whether the advantages (e.g. protecting homes and schools) are worth the cost (the loss of farmland further along the coast).

Use the evidence to make some inferences about the possible impacts of coastal management at Withernsea on coastal processes further along the coast.

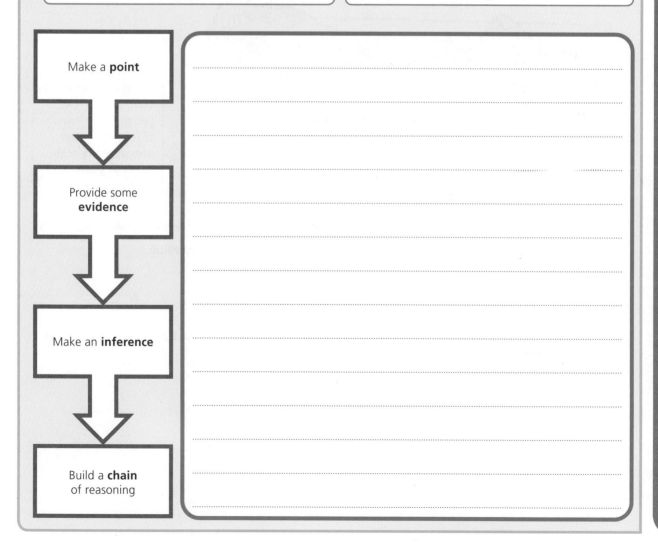

Make a **point**

Provide some **evidence**

Make an **inference**

Build a **chain** of reasoning

# Managed retreat

We cannot prevent erosion or flooding on every coastline. Coastal management is too expensive. If the costs are greater than the benefits then the coastline is allowed to retreat in a controlled and managed way. This is known as **managed retreat** or **managed realignment**.

**6** Which key geography term is being described here? <u>Underline</u> **one** term.

**Definition:** The strip of coast that is between high tide and low tide.

beach          rock pools          intertidal zone

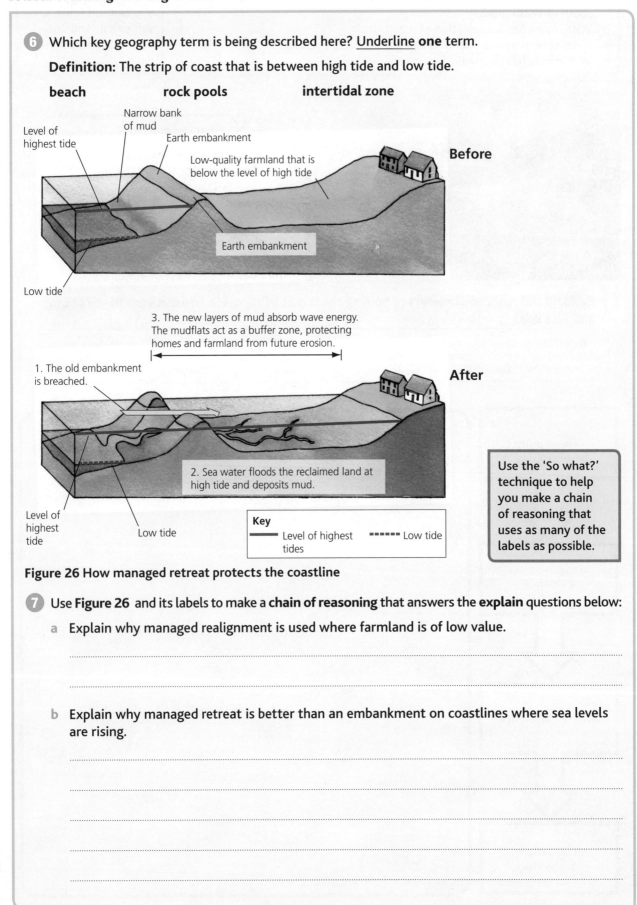

**Figure 26** How managed retreat protects the coastline

> Use the 'So what?' technique to help you make a chain of reasoning that uses as many of the labels as possible.

**7** Use **Figure 26** and its labels to make a **chain of reasoning** that answers the **explain** questions below:

a  Explain why managed realignment is used where farmland is of low value.

................................................................................................................................

................................................................................................................................

b  Explain why managed retreat is better than an embankment on coastlines where sea levels are rising.

................................................................................................................................

................................................................................................................................

................................................................................................................................

................................................................................................................................

................................................................................................................................

# Sea level rise and flood risk

Some coastlines will be at greater risk in the **future** because of **sea level rise**.
About 65 million people live in **Small Island Developing States (SIDs)**. **SIDs**
are vulnerable to **coastal flooding** and **climate change**. Sea level rises may force
people to move from their homes. They will become **environmental refugees**.

You must use evidence in the photo – in this case the concrete pillars and the shutters over the windows give important clues. What can you infer?

Use the 'What might...?' or 'What could...?' questions.

Remember that tropical storms create strong winds, large waves and a temporary rise in sea levels known as a storm surge.

**Figure 27 A house built at the top of the beach in Tobago**

## Fact Box

- Tobago is a small Caribbean island. It is part of Trinidad and Tobago which is a Small Island State.
- At present there is a one in 25 chance each year of Tobago having a hurricane.

**8** 'These homes are well protected from **future** coastal hazards.' To what extent do you agree? Use evidence in the Fact Box and photo.

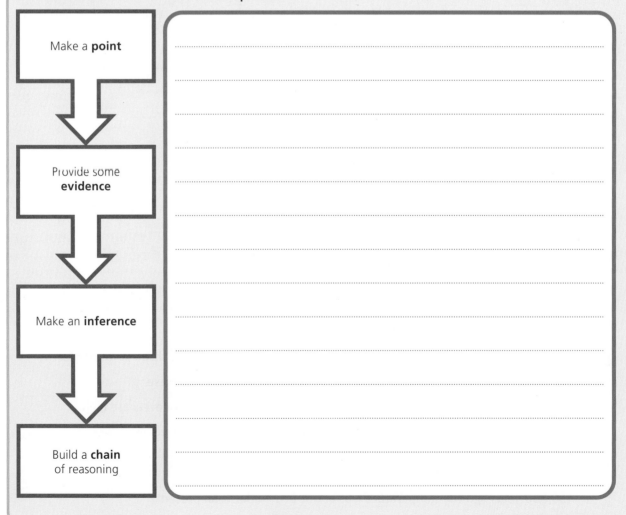

Make a **point**

Provide some **evidence**

Make an **inference**

Build a **chain** of reasoning

## Theme 5 Weather, climate and ecosystems

Theme 5 is a core theme so you **must** revise it. It is examined in Paper 2, Question 1.

### Climate change

The climate has changed over the last 2.6 million years – the **Quaternary Period**. Sometimes the climate has been cold with ice sheets and glaciers. These are **glacial periods**. At other times the ice has melted. These are **interglacial** periods. Evidence of climate change comes from:

- **ice cores** (which contain $CO_2$ trapped in the ice)
- **temperature records** (since about 1850)
- **tree rings** (because rings are thicker when the climate is warmer and wetter).

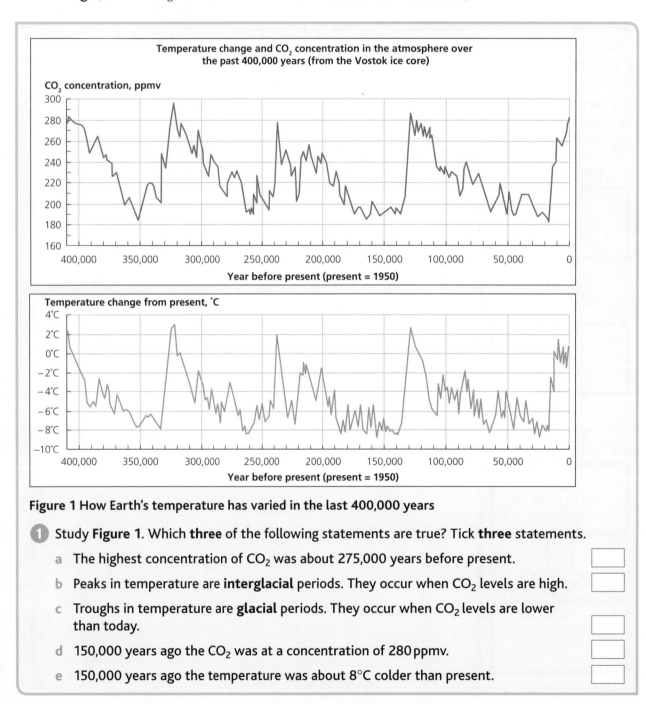

**Figure 1** How Earth's temperature has varied in the last 400,000 years

1 Study **Figure 1**. Which **three** of the following statements are true? Tick **three** statements.

a The highest concentration of $CO_2$ was about 275,000 years before present. ☐

b Peaks in temperature are **interglacial** periods. They occur when $CO_2$ levels are high. ☐

c Troughs in temperature are **glacial** periods. They occur when $CO_2$ levels are lower than today. ☐

d 150,000 years ago the $CO_2$ was at a concentration of 280 ppmv. ☐

e 150,000 years ago the temperature was about 8°C colder than present. ☐

## Possible causes of climate change

Some causes of climate change are natural. Factors such as changes in the **Earth's orbit**, **volcanic eruptions** and variations in solar output caused climate to change over the Quaternary Period. Human activities such as **burning fossil fuels**, agriculture and **deforestation** change the amount of carbon dioxide in the atmosphere. This causes more climate change.

**2** Which key geography term is being described here? <u>Underline</u> **one** term.

**Definition:** The way in which certain gases, including carbon dioxide, trap heat in the atmosphere.

**global cooling**        **climate change**        **greenhouse effect**

**3** Join the **Points** to the **Explanations** with an arrow to show that you understand the causes of climate change.

| Points | Explanations |
|---|---|
| The path taken by the Earth as it travels around the Sun varies... | ...so solar energy is blocked as it passes through the atmosphere which causes global cooling. |
| Volcanoes emit $SO_2$ when they erupt... | ...so methane is able to act as a powerful greenhouse gas which leads to climate change. |
| When trees are chopped down and burnt they release $CO_2$ into the atmosphere... | ...so the amount of solar energy received by the Earth varies over a 96,000 year cycle. |
| Rice farming, dairy farms and beef farms emit methane as a waste product... | ...so extra $CO_2$ traps heat in the atmosphere. This is the enhanced greenhouse effect. |

## Circulation of the atmosphere

The movement of air in our atmosphere is driven by the Sun warming the Earth. The Equatorial regions are the hottest part of the Earth because the Sun is overhead. Excess heat at the Equator triggers **atmospheric circulation**, causing warm air to spread outwards towards the Tropics and creating **trade winds**.

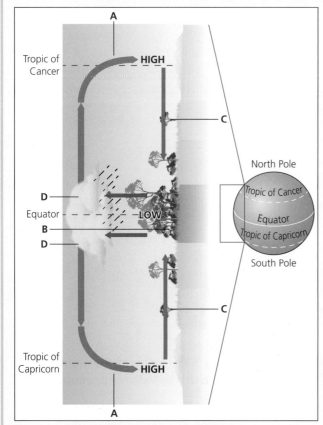

**Figure 2 Circulation of the atmosphere**

**4** Match the letters on **Figure 2** to the descriptions below.

| Letter | Description |
|---|---|
| | The Earth heats the air above it. The air rises, creating an area of low pressure. |
| | The rising air reaches a boundary in the atmosphere called the tropopause. The air flows towards the North and South Poles, creating winds high in the atmosphere. |
| | At about 30°N and 30°S the air sinks, creating an area of high pressure. |
| | The air is pulled back towards the Equator by the area of low pressure. This creates low-level winds in the atmosphere called the trade winds. |

# Low pressure weather hazards

**Low pressure** in the atmosphere is caused by air rising. Low pressure creates weather hazards such as **monsoon rains** and **tropical storms**. Tropical storms begin when air is warmed over warm oceans (the water has to be over 27°C for several weeks).

## Example of a low pressure hazard

You will have studied an example of a low pressure hazard such as a tropical storm or monsoon. You need to remember when it happened, the places it affected, how they were affected and how people responded to this hazard. Use the table below to summarise your example.

| When it happened | |
|---|---|
| Place(s) affected | |

| Effects on people/the economy | Effects on the environment |
|---|---|
| | |

| How people responded, e.g. emergency aid or repairs to flood defences | |
|---|---|

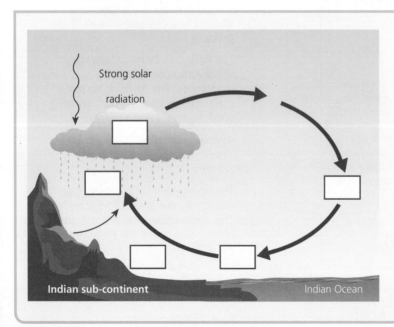

**Figure 3 The cause of the monsoon rains over the Indian sub-continent**

60

5. Study **Figure 3** on page 60 and the labels below. Complete the diagram by adding the letters A–E to the correct boxes on **Figure 3**.

  A  The ground is heated by energy from the Sun, which is directly overhead.

  B  Air, which is full of moisture, rises creating low pressure over the Indian sub-continent.

  C  Air sinks creating high pressure over the Indian Ocean.

  D  Moist air over the Indian Ocean is drawn towards the area of low pressure.

  E  Moisture in the air condenses forming towering rain clouds.

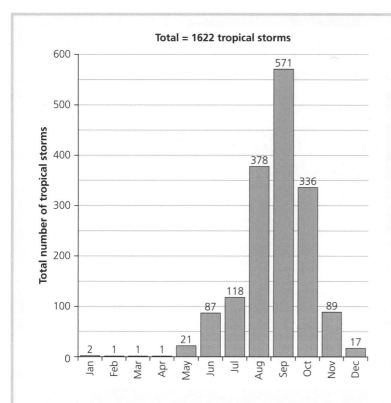

**Figure 4** Total number of Atlantic/ Caribbean tropical storms (1851–2015)

6. Study **Figure 4**. What percentage of storms occurs in the period August to October? Show your working to 2 dp.

**Total = 1622 tropical storms**

# Effects of tropical storms

Tropical storms create very **strong winds** and **heavy rain** that can cause **landslides**. The low air pressure causes a temporary rise in sea levels known as a **storm surge** which may **flood** coastal areas.

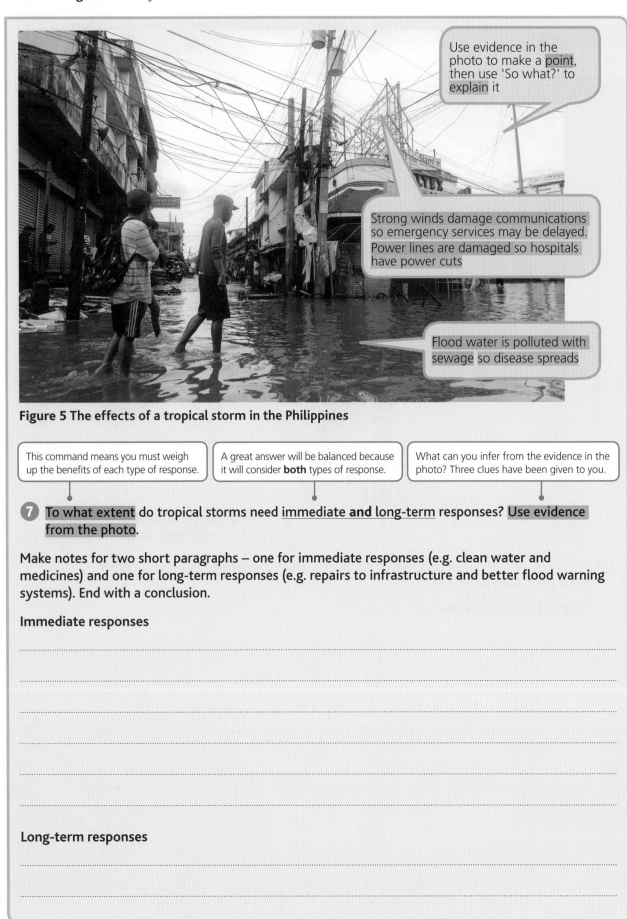

**Figure 5 The effects of a tropical storm in the Philippines**

Use evidence in the photo to make a point, then use 'So what?' to explain it

Strong winds damage communications so emergency services may be delayed. Power lines are damaged so hospitals have power cuts

Flood water is polluted with sewage so disease spreads

This command means you must weigh up the benefits of each type of response.

A great answer will be balanced because it will consider **both** types of response.

What can you infer from the evidence in the photo? Three clues have been given to you.

**7** **To what extent** do tropical storms need immediate **and** long-term responses? Use evidence from the photo.

Make notes for two short paragraphs – one for immediate responses (e.g. clean water and medicines) and one for long-term responses (e.g. repairs to infrastructure and better flood warning systems). End with a conclusion.

**Immediate responses**

...................................................................................................................................

...................................................................................................................................

...................................................................................................................................

...................................................................................................................................

...................................................................................................................................

...................................................................................................................................

**Long-term responses**

...................................................................................................................................

...................................................................................................................................

........................................................................................

........................................................................................

........................................................................................

........................................................................................

**Conclusion**

........................................................................................

........................................................................................

........................................................................................

........................................................................................

........................................................................................

## High pressure weather hazards

**High pressure** in the atmosphere is caused by air sinking. **Droughts** and **heatwaves** are caused by **high pressure** becoming stuck over a region for weeks or months. The climate is changing and droughts seem to be getting more **severe** and more **frequent** in some parts of the world.

### Example of a high pressure hazard

You will have studied an example of a high pressure hazard such as a drought or heatwave. You need to remember when it happened, the places it affected, how they were affected and how people responded to this hazard. Use the table below to summarise your example.

| When it happened | |
|---|---|
| Place(s) affected | |

| Effects on people/the economy | Effects on the environment |
|---|---|
| | |

| How people responded, e.g. emergency aid or water rationing | |
|---|---|
| | |

# Ecosystems

Plants in an ecosystem get their energy from the Sun in a process called **photosynthesis**. They use this energy to produce food in the form of leaves, seeds or fruit – so plants are **producers** in the ecosystem. In a UK woodland, the leaves, seeds and fruit are eaten by caterpillars, insects, birds or mice. These are **consumers**. Energy is recycled in an ecosystem through **nutrient cycling** – see **Figure 6**.

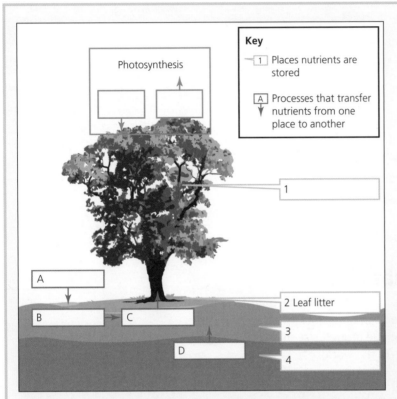

**Figure 6 Nutrient cycling in a UK woodland**

8  Complete the orange boxes in **Figure 6** to show the gases transferred by photosynthesis.

9  Match the following explanations to the correct letter or number on the diagram to explain nutrient cycling.

| Letter/number | Explanation |
|---|---|
| | Nutrients are stored in rocks. |
| | Weathering breaks down rocks and releases nutrients into the soil. |
| | Nutrients are stored in the soil. |
| | Nutrients are stored in the cells of the tree. |
| | Dead leaves and branches fall from the tree. |
| | Nutrients are stored temporarily in the dead leaves and branches on the forest floor. |
| | Beetles and earthworms break down the dead leaves. Bacteria and fungi (decomposers) release the nutrients into the soil. |
| | Water dissolves the nutrients. The tree takes in the water and nutrients through its roots. |

## Example of a small UK ecosystem

You will have studied an example of a small-scale UK ecosystem such as a woodland, hedgerow or sand dune. Use the table below to summarise key facts and figures of your example.

| | |
|---|---|
| Type of ecosystem | |
| Example of a food chain | |
| How people have affected the ecosystem | |

# Rainforest ecosystems

Tropical rainforests are an example of a **large-scale** ecosystem or **biome**. Most tropical rainforests grow within 10 degrees of the Equator. The climate is:

- hot every month; it is usually between 23°C and 30°C
- wet for most months of the year – total rainfall is between 2000 mm and 4000 mm per year.

Plants that grow in tropical rainforests are specially adapted to this climate.

**10** a Circle the correct word in each description below.

b Match the descriptions to the labels on **Figure 7**.

| Label | How plants are adapted |
|---|---|
| | Leaves are *waxy / porous* and have a drip-tip so that rainwater runs off quickly, preventing rot. |
| | Plants grow *slowly / quickly* to reach the *nutrients / sunlight* above. |
| | The forest floor is *shaded / brightly lit* so plants on the forest floor have *small / large* leaves so they can catch enough sunlight. |
| | The upper leaves form an umbrella-like *canopy / emergent* layer. This prevents a lot of light reaching the forest floor. |

**Figure 7 Tropical rainforest plants**

# Managing rainforests

Tropical rainforests contain a lot of useful resources, including timber, minerals (e.g. iron ore) and land for **agriculture**. Rainforests are also cut down to make space for energy schemes (e.g. hydro-electricity projects), new roads and new towns. Cutting down trees is called **deforestation**. Deforestation creates wealth (an economic impact), but has negative impacts for wildlife (an environmental impact).

**11** Explain why deforestation leads to positive and negative impacts by doing the following:

a Linking the **Points** about deforestation on the left to the **Explanations** on the right. One example has been done for you.

b Use a highlighter to show how each explanation has been extended further.

c Which of these impacts is negative and which is positive?

Positive impacts .............................................................. Negative impacts ..............................................................

| Points | Explanations |
|---|---|
| **1** The tropical rainforest stores a lot of carbon … | … so tropical countries are able to create wealth by exporting food crops leading to a better balance of trade. |
| **2** When trees are cut down the canopy is destroyed … | … so when it is burned CO$_2$ is released into the atmosphere causing climate change. |
| **3** Large firms create commercial farms growing palm oil or oranges … | … so interception is reduced leading to soil erosion and a reduction in soil fertility. |
| **4** Large areas of rainforest are replaced by farms growing one crop, such as soy … | … so the rainforest becomes more accessible, which means that more people move in looking for work. |
| **5** Deforestation allows minerals such as iron ore to be extracted … | … so the rich variety of plants is replaced by just one species of plant, which means much less biodiversity. |
| **6** Roads are built through the forest to remove logs … | … so well-paid jobs are created in mining and engineering, which leads to a demand for better training and research. |

> The 'So what?' technique has been used to explain each point (highlighted in yellow). Notice how the sentence has been extended again to add further explanation (highlighted in green).

> Removing the canopy has exposed soil to rainfall. What can you infer from this? Ask yourself 'What might?' or 'What ought?' questions.

**Figure 8 Deforestation of the tropical rainforest in Papua New Guinea**

Use what you learned in class about rainforests to link deforestation to negative impacts like soil erosion or positive impacts like job creation.

Make sure you consider at least one negative impact on the environment (e.g. on biodiversity). To get some balance in your answer, you should also consider any positive impacts.

**12** 'Deforestation has so many negative impacts it ought to be stopped.' To what extent do you agree? Use evidence in the photo.

Refer to evidence in the photo such as the loss of the canopy and how this may affect soils and wildlife.

You must weigh up the evidence and make a decision. Can you think of any bad impacts on the economy?

Make a **point**

Provide some **evidence**

Make an **inference**

Build a **chain** of reasoning

# Theme 6 Development and resource issues

Theme 6 is a core theme so you **must** revise it. It is examined in Paper 2, Question 2.

## Measuring global inequality

If we put countries in rank order by their wealth from richest to poorest, we create a **continuum** of **economic development**. The economic wealth of individual countries grows at different rates. Some **newly industrialised countries (NICs)** have become much wealthier in recent years – so the **development gap** between richer and poorer countries is **dynamic** – it is constantly changing.

> **1** Which key geography term is being described here? <u>Underline</u> **one** term.
>
> **Definition:** The economic difference between richer and poorer countries.
>
> development gap          trade bloc          continuum of development

## Trade

Countries are connected to each other through **trade** – the buying and selling of **goods** and **services**.

> **2** Match **six** key terms to the correct definition below. **Two** terms are not needed.
>
> exports      imports      fairtrade        trade bloc      tariffs
>
> subsidies      multi-national companies        quotas
>
> | Term | Definition |
> | --- | --- |
> | | Taxes that are paid on goods when they are brought into a country. |
> | | Limits on the amount of goods that can be bought. |
> | | Goods and services that are sold by a country. |
> | | Goods and services that are bought by a country. |
> | | A group of countries that trades freely with each other. |
> | | A method of trade that ensures producers get reasonable pay. |

## Multi-national companies (MNCs)

**Multi-national companies (MNCs)** are large businesses that have **branches** (factories, research facilities and offices) in more than one country. They provide **advantages** for the **host** countries, for example, they create employment. They also create **disadvantages**, for example, at times of recession the MNC may close branches in host countries to protect the main jobs in the home country.

> **Figure 9 Advantages and disadvantages of MNCs**
>
> | | | |
> | --- | --- | --- |
> | 1 Jobs are created at various levels of skill. | 5 Wages are often lower in the host country. | 9 Workers are trained and learn new industrial or business skills. |
> | 2 Local taxes may be lower than in the country where the MNC has its head office. | 6 Workers spend money in local shops. | 10 Environmental laws (e.g. about waste) may be less strong in the host country. |
> | 3 NICs have large populations who are potential customers. | 7 The most highly paid jobs (e.g. research) stay in the country where the MNC has its head office. | 11 The MNC pays local taxes. |
> | 4 There may be no minimum wage. | 8 The local government is able to spend tax revenue. | 12 Other local firms get contracts, e.g. firms that supply component parts. |

3 Study **Figure 9**. Write the number of each of the twelve statements into the correct place in the following table. **One** has been done for you.

|  | Advantages | Disadvantages |
|---|---|---|
| For the MNC |  |  |
| For the host country | 1 |  |

**Figure 10** The location of Toyota's branches outside of Japan

4 Study **Figure 10**.

a How many vehicles are produced each year in Brazil?

.............................................................

b Describe the distribution of Toyota's branches. Use map evidence only.

...................................................................................................................................

...................................................................................................................................

...................................................................................................................................

> Try to give an **overview** of the whole map rather than listing place names. For example, use comparative phrases such as 'There are significantly more branches in Asia and North America than in Europe or Africa.' **Never** use your own knowledge in a question that asks you to use map evidence. For example, you may know that Brazil is an NIC, but don't use this information.

5 Use the map in **Figure 10** and the phrases in **Figure 9** to make a **chain of reasoning** that answers the **explain** question below:

Explain why MNCs like Toyota have factories in several NICs.

...................................................................................................................................

...................................................................................................................................

...................................................................................................................................

...................................................................................................................................

...................................................................................................................................

# Tourism

The growth of the global tourist industry has created wealth in many countries. **Tourism** creates jobs **directly** in new airports and hotels. Tourism also creates extra work **indirectly** for people like farmers and fishermen who supply food to the hotels. However, jobs created by tourism may be **seasonal** or badly paid. Large hotels are owned by multi-national companies (**MNCs**) so some profits are lost. Tourists may stop coming if the country has a natural disaster or is affected by terrorism.

6 Which key geography term is being described here? <u>Underline</u> **one** term.

**Definition:** The proportion of people employed in tourism compared to other jobs such as farming.

employment structure          indirect jobs          seasonal employment

7 Study the phrases below.

Faster long-haul flights

More fuel-efficient aircraft engines

More TV channels

Use of apps like Airbnb

Travel channels

Growing mobile phone ownership

Faster download of data on mobiles

3G and 4G networks

Use of mobiles and cards for payments

Translation apps

Use the phrases above to make **two chains of reasoning** that answer the **understanding** question below:

Give **two** reasons for the growth of global tourism.

.....................................................................................................................................................................

.....................................................................................................................................................................

.....................................................................................................................................................................

.....................................................................................................................................................................

## Example of tourism

You will have studied examples of how the growth of tourism in an LIC or NIC can have positive and negative impacts. Use the table below to summarise **one** example.

| | Name of country: ......................................................... | |
|---|---|---|
| | Positive effects | Negative effects |
| Employment | | |
| Environment | | |
| Culture | | |
| Infrastructure | | |

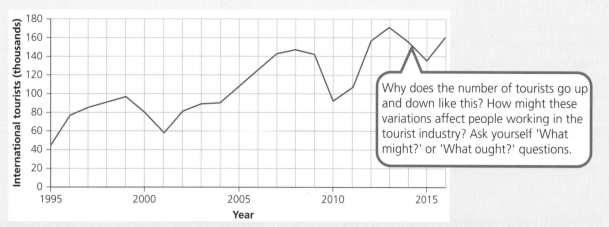

**Figure 10 The number of international tourists visiting the Gambia, West Africa**

> Why does the number of tourists go up and down like this? How might these variations affect people working in the tourist industry? Ask yourself 'What might?' or 'What ought?' questions.

**Figure 11 A cruise ship visiting Tobago, the Caribbean**

> A lot of tourists visit the Caribbean on cruise ships. What can you infer from this? Ask yourself 'What might' or 'What ought' questions.

> You must show you understand how tourism can create jobs, reduce poverty and increase wealth.

**8** 'The growth of tourism is able to reduce the development gap.' **To what extent do you agree?** Use evidence in the graph and photo.

> You must show you can weigh up the evidence and reach a judgement.

Answer this question in three short paragraphs. First, create an argument. Second, a counter-argument. Finally, use the 'washing line' technique.

On the one hand …

........................................................................

........................................................................

........................................................................

........................................................................

> Use **PEIC** to make an argument that tourism creates jobs – you could mention direct and indirect jobs.

On the other hand …

........................................................................

........................................................................

........................................................................

........................................................................

> Create a counter-argument. Using evidence from the graph or photo, show why tourism can fail to reduce the development gap.

I fully/largely/partially agree/disagree …

........................................................................

........................................................................

........................................................................

> **Link** back to the question. State whether tourism is effective at reducing the development gap or not. Use the 'washing line' technique.

# Water resources

The global **demand** for water is increasing. As the world's population grows, we need more water for drinking and **sanitation** (flushing toilets and treating sewage) as well as for industry and agriculture. There are various ways of making sure that everyone has enough water. We can:

- **transfer water** from one place to another – sometimes across national boundaries
- use **small-scale** schemes such as collecting rainwater from the roof of buildings
- **abstract** (take) water from the **ground**.

**9** Which key geography term is being described here? <u>Underline</u> **one** term.

**Definition:** When everyone has clean water and sanitation and the economy has enough water to grow food and make things.

**water transfer**      **water security**      **ecological footprint**

**10** Match **five** key terms to the correct definition below. **Two** terms are not needed.

**irrigation**      **rainwater harvesting**      **water footprint**

**abstraction**      **water transfer scheme**      **aquifer**      **over-abstraction**

| Term | Definition |
|---|---|
| | The amount of water used to produce an item of food or make a product such as an item of clothing. |
| | The process of spreading water across fields to help crops grow. |
| | The process of taking water from a river or from the ground. |
| | When water is taken from the ground quicker than it can be replaced by rainwater soaking down into the ground |
| | The use of rivers and pipes to move water from a place where there is plenty of water to a place that does not have enough. |

> How is rainwater being collected and stored? What can you infer from this? Ask yourself 'What might?' or 'What ought?' questions.

**Figure 12 Rainwater harvesting on a house in Sri Lanka**

**Figure 13 Advantages and disadvantages of large-scale water transfer schemes and small-scale rainwater harvesting**

| | | |
|---|---|---|
| 1 Provides water for irrigating large commercial farms. | 2 Uses cheap materials. | 3 Difficult to store enough water to survive a long drought. |
| 4 Can take many years to construct. | 5 Provides water for growing vegetables in a small garden. | 6 Provides large quantities of clean water for homes. |
| 7 Does not need special training to install or maintain. | 8 Expensive so the government may need to borrow money to pay for its construction. | 9 Needs complex agreement if the river crosses national borders. |
| 10 May also be used to generate cheap HEP to use in industry. | 11 Very easy to fix if it goes wrong. | 12 May flood valuable farmland and displace people. |

**11** Study **Figure 13**. Write the number of each of the twelve statements at the bottom of page 72 into the correct place in the following table. **One** has been done for you.

|  | Advantages | Disadvantages |
|---|---|---|
| Large-scale water transfer | 1 |  |
| Rainwater harvesting |  |  |

**12** Use **Figure 12** and the phrases in **Figure 13** to make a **chain of reasoning** that answers the **explain** question below:

Explain why rainwater harvesting is described as a sustainable technology.

> Think about the advantages of rainwater harvesting and use these to create a chain of reasoning that explains why this technology is sustainable.

...................................................................................................................

...................................................................................................................

...................................................................................................................

## Over-abstraction of groundwater

In some places water is taken from the ground quicker than it can be **recharged** by rainwater so the level of the **water table** drops. One such place is Gujarat in north-west India.

**13** Explain why so much groundwater is abstracted in Gujarat by linking the **Points** on the left with an arrow to the **Explanations** on the right. One has been done for you.

| Points | Explanations |
|---|---|
| States in North West India have a long dry season when there is little rainfall... | ...so people use groundwater because they think it is cleaner than water from surface stores. |
| Lakes and rivers are often polluted by human waste... | ...so drilling a well for groundwater is seen as a cheaper option for people in cities. |
| Modern farms use techniques learned in the 'green revolution' to grow more food... | ...so lakes and rivers can be dry for long periods of time so people rely on groundwater instead. |
| In many urban slums people have to buy water in jerry cans that are very expensive... | ...so farms use more water because new crop varieties need more water than traditional crops. |
| The amount of rain that falls during the monsoon varies from year to year... | ...so wells get deeper and the level of the water table falls even further as water is abstracted. |
| Cheap electricity means farmers can abstract water from deeper wells... | ...so in years with poor rainfall more water is taken from the ground than goes back naturally. |

**14** Use some of the ideas in the previous activity to make a **chain of reasoning** that answers the **explain** question below.

Explain why some farmers are worried about what will happen to water supplies in the future.

...................................................................................................................

...................................................................................................................

...................................................................................................................

# Regional economic development

Countries often develop with one or more regions that are wealthier than others. There may be economic, social, cultural, political or environmental reasons for these **regional inequalities**. Poverty and migration are consequences of regional inequality.

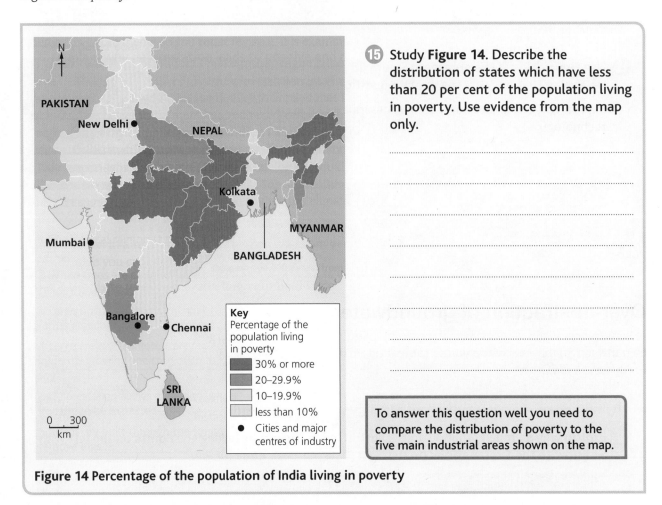

**Figure 14 Percentage of the population of India living in poverty**

15 Study **Figure 14**. Describe the distribution of states which have less than 20 per cent of the population living in poverty. Use evidence from the map only.

.................................................................

.................................................................

.................................................................

.................................................................

.................................................................

.................................................................

.................................................................

.................................................................

To answer this question well you need to compare the distribution of poverty to the five main industrial areas shown on the map.

## Example of regional inequality

You will have studied one LIC or NIC that has regional inequalities. Use the table below to summarise your example.

| Name of country | |
|---|---|
| Reasons for inequality (e.g. economic, social, cultural, political or environmental) | |

| Social consequences | Economic consequences |
|---|---|
| | |

# Regional inequality in the UK

In the UK there is a **North–South divide**. The south-east region of the UK attracts a lot of **high-tech** companies. They benefit from being close to major airports and financial services in the City of London. Economic growth attracts other businesses and migrant workers – a concept called the **positive multiplier**. In **deprived** regions of the UK the closure of businesses leads to a **negative multiplier**.

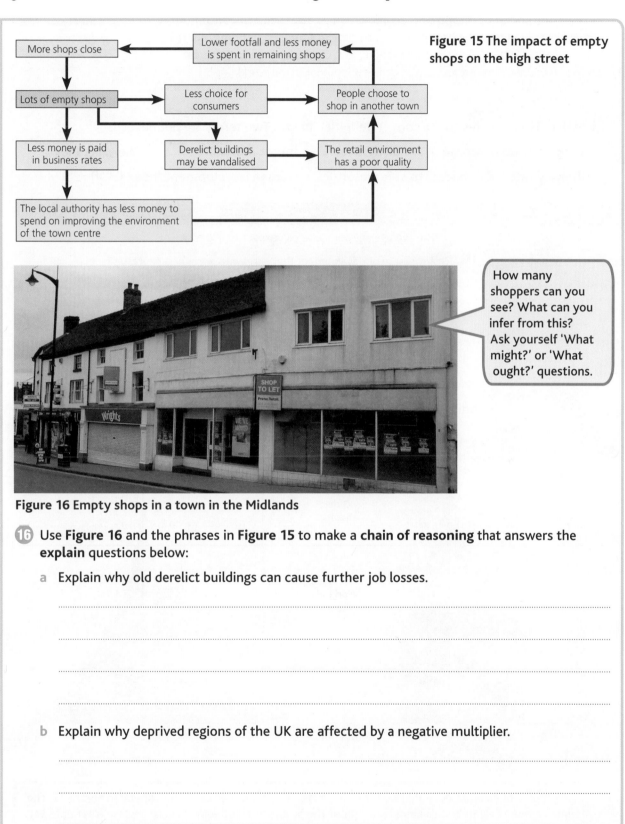

**Figure 15 The impact of empty shops on the high street**

How many shoppers can you see? What can you infer from this? Ask yourself 'What might?' or 'What ought?' questions.

**Figure 16 Empty shops in a town in the Midlands**

16  Use **Figure 16** and the phrases in **Figure 15** to make a **chain of reasoning** that answers the **explain** questions below:

a   Explain why old derelict buildings can cause further job losses.

..................................................................................................................................................

..................................................................................................................................................

..................................................................................................................................................

..................................................................................................................................................

b   Explain why deprived regions of the UK are affected by a negative multiplier.

..................................................................................................................................................

..................................................................................................................................................

..................................................................................................................................................

..................................................................................................................................................

# Theme 7 Social development

Theme 7 is an optional theme. It is examined in Paper 2, Question 3.

Tick the box if you studied this theme. ☐

## Measuring social development

There are a number of different ways that data about **gender** and **health** can be used to measure **social development**. Countries can be arranged in a **continuum** of social development from the healthiest to unhealthiest countries or from countries that have equality for the genders to those that do not. Countries in **sub-Saharan Africa** tend to have the lowest levels of social development in the world.

---

**1** Match **five** measures to the correct definition below. **Two** terms are not needed.

**Gross National Income**   **birth rate**   **infant mortality**   **life expectancy**

**literacy rate**   **access to safe water**   **Human Development Index (HDI)**

| Term | Definition |
|---|---|
| | The average age someone can expect to live to. |
| | A measure of development that takes into account a country's level of education, its wealth and average life expectancy. |
| | The number of children born in one year for every 1,000 people in a country's population. |
| | The number of children who die before the age of one for every 1,000 that are born. |
| | The percentage of people, aged 15 and over, who can read and write. |

### India 2017

| | Male | Female | |
|---|---|---|---|
| 100+ | 0.0% | 0.0% | |
| 95–99 | 0.0% | 0.0% | |
| 90–94 | 0.0% | 0.0% | |
| 85–89 | 0.1% | 0.1% | |
| 80–84 | 0.3% | 0.3% | Older adults |
| 75–79 | 0.5% | 0.6% | |
| 70–74 | 0.8% | 0.8% | |
| 65–69 | 1.2% | 1.2% | |
| 60–64 | 1.7% | 1.7% | |
| 55–59 | 2.1% | 2.0% | |
| 50–54 | 2.5% | 2.4% | |
| 45–49 | 2.8% | 2.7% | Working adults |
| 40–44 | 3.2% | 3.0% | |
| 35–39 | 3.7% | 3.5% | |
| 30–34 | 4.2% | 3.9% | |
| 25–29 | 4.5% | 4.1% | |
| 20–24 | 4.7% | 4.2% | |
| 15–19 | 4.9% | 4.4% | |
| 10–14 | 5.0% | 4.5% | Children |
| 5–9 | 4.9% | 4.4% | |
| 0–4 | 4.8% | 4.3% | |

### Gambia 2017

| | Male | Female |
|---|---|---|
| 100+ | 0.0% | 0.0% |
| 95–99 | 0.0% | 0.0% |
| 90–94 | 0.0% | 0.0% |
| 85–89 | 0.0% | 0.0% |
| 80–84 | 0.1% | 0.1% |
| 75–79 | 0.2% | 0.2% |
| 70–74 | 0.4% | 0.3% |
| 65–69 | 0.5% | 0.4% |
| 60–64 | 0.7% | 0.7% |
| 55–59 | 0.9% | 1.0% |
| 50–54 | 1.2% | 1.3% |
| 45–49 | 1.5% | 1.7% |
| 40–44 | 1.8% | 2.1% |
| 35–39 | 2.5% | 2.8% |
| 30–34 | 3.0% | 3.3% |
| 25–29 | 3.6% | 3.8% |
| 20–24 | 4.4% | 4.5% |
| 15–19 | 5.4% | 5.4% |
| 10–14 | 6.4% | 6.3% |
| 5–9 | 7.7% | 7.5% |
| 0–4 | 9.2% | 9.0% |

**Figure 17** Population pyramid for India (an NIC) and Gambia (an LIC) in sub-Saharan Africa

Always divide a population pyramid into three sections. The author has done this for you in **Figure 17**. The examiners are unlikely to be this kind! Think about the proportion of people in each sector. What does this tell you about the birth rate? Are enough people working to look after the older adults? Also, are there the same numbers of men as women? If not, why not? What can you infer?

**2** Describe **two** main differences between the population structure of Gambia and India. Use **Figure 17**. Go on to suggest what can be inferred from this. One has been done for you.

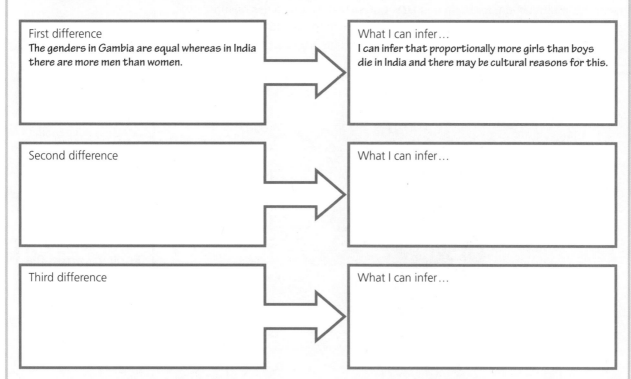

First difference
The genders in Gambia are equal whereas in India there are more men than women.

What I can infer…
I can infer that proportionally more girls than boys die in India and there may be cultural reasons for this.

Second difference

What I can infer…

Third difference

What I can infer…

**3** **a** Population pyramids change shape over time as a country become more economically developed. Explain the reasons for the changing population structure by linking the **Points** on the left to the **Explanations** on the right. One example has been done for you.

**b** Use a highlighter to show how each explanation has been extended further.

| Points | Explanations |
| --- | --- |
| The status of women in most South Asian countries is improving so women have a more equal place in society… | …so people live longer and have fewer children which means that the population stops growing. |
| Manufacturing and service industries are replacing agriculture as the main employment in many parts of South Asia… | …so education for girls improves and better jobs become available so women choose to have fewer children. |
| In South Asia healthcare is improving and many women choose to start their family later in life… | …so deaths from cholera and typhoid are falling so death rates are falling. |
| In many countries of sub-Saharan Africa pay is low and there are no social benefits or pensions for poor people… | …so fewer people work on farms and there is much less need for children to work so the birth rate falls. |
| In most countries of sub-Saharan Africa access to clean water is improving (although not everyone has access to it)… | …so infant mortality is falling which means that the death rate also falls. |
| The use of mosquito nets in many countries of sub-Saharan Africa prevents the spread of malaria amongst vulnerable people… | …so parents rely on their children to help the family earn a living so birth rates remain high. |

The 'So what?' technique has been used to explain each point (and highlighted in yellow). Notice how the sentence has been extended again to add further explanation (highlighted in green).

# Healthcare issues

Countries in sub-Saharan Africa have higher rates of **infant mortality** than countries in South Asia. The causes are complex but include issues such as water insecurity, malnutrition, **malaria** and **HIV**. Consequently, countries of sub-Saharan Africa have some of the lowest **HDI** scores in the world.

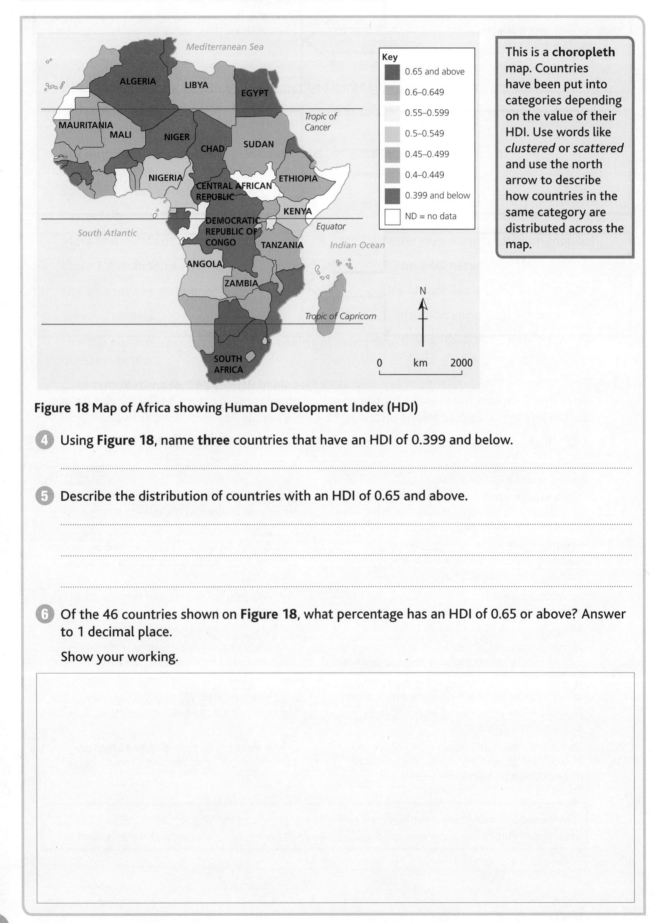

**Figure 18** Map of Africa showing Human Development Index (HDI)

4 Using **Figure 18**, name **three** countries that have an HDI of 0.399 and below.

.......................................................................................................................................................

5 Describe the distribution of countries with an HDI of 0.65 and above.

.......................................................................................................................................................

.......................................................................................................................................................

.......................................................................................................................................................

6 Of the 46 countries shown on **Figure 18**, what percentage has an HDI of 0.65 or above? Answer to 1 decimal place.

Show your working.

# Child labour

Not every child in South Asia and sub-Saharan Africa goes to school. Many children help the family by caring for younger children, collecting water or working. Some work in the **informal sector** – jobs such as street vending or cleaning shoes. Other children work in factories, farms or quarries.

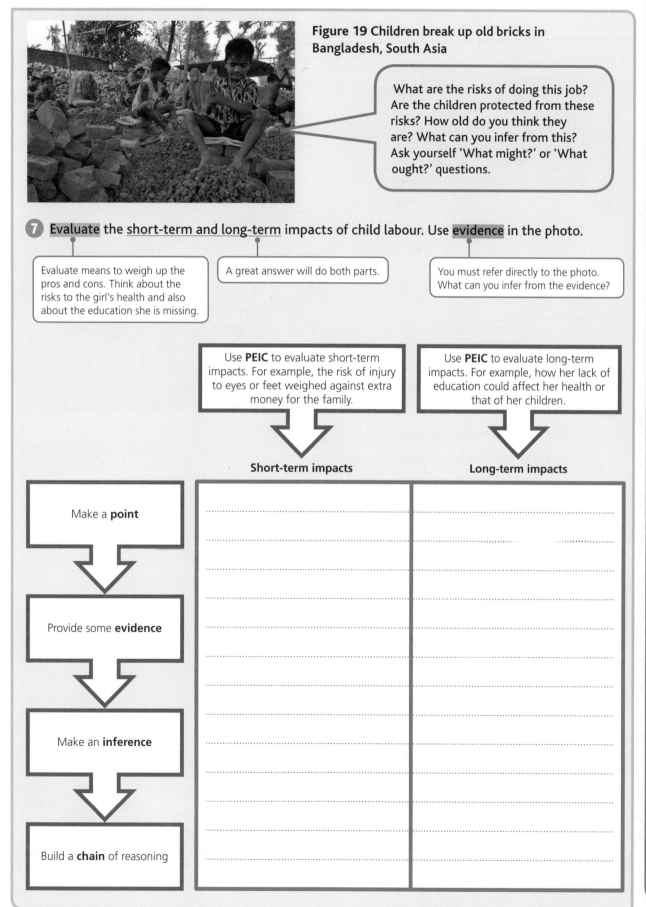

**Figure 19 Children break up old bricks in Bangladesh, South Asia**

What are the risks of doing this job? Are the children protected from these risks? How old do you think they are? What can you infer from this? Ask yourself 'What might?' or 'What ought?' questions.

7 **Evaluate** the <u>short-term and long-term</u> impacts of child labour. Use evidence in the photo.

Evaluate means to weigh up the pros and cons. Think about the risks to the girl's health and also about the education she is missing.

A great answer will do both parts.

You must refer directly to the photo. What can you infer from the evidence?

Use **PEIC** to evaluate short-term impacts. For example, the risk of injury to eyes or feet weighed against extra money for the family.

Use **PEIC** to evaluate long-term impacts. For example, how her lack of education could affect her health or that of her children.

Short-term impacts

Long-term impacts

Make a **point**

Provide some **evidence**

Make an **inference**

Build a **chain** of reasoning

79

# Education of girls

In many countries of South Asia and sub-Saharan Africa adult literacy is lower in women than in men. This difference between men and women is an example of **gender inequality**. This is because many girls do not complete their education.

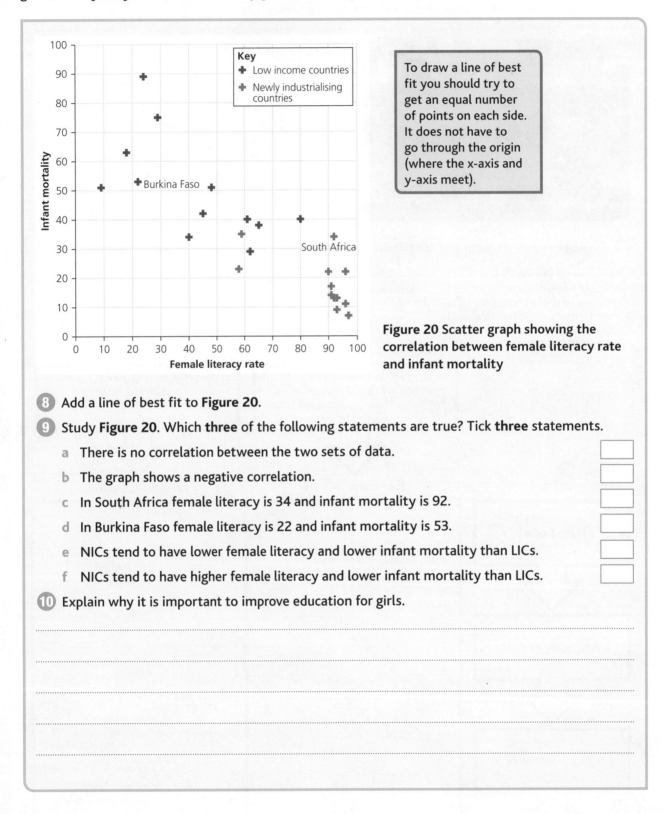

To draw a line of best fit you should try to get an equal number of points on each side. It does not have to go through the origin (where the x-axis and y-axis meet).

**Figure 20** Scatter graph showing the correlation between female literacy rate and infant mortality

8 Add a line of best fit to **Figure 20**.

9 Study **Figure 20**. Which **three** of the following statements are true? Tick **three** statements.

a There is no correlation between the two sets of data. ☐

b The graph shows a negative correlation. ☐

c In South Africa female literacy is 34 and infant mortality is 92. ☐

d In Burkina Faso female literacy is 22 and infant mortality is 53. ☐

e NICs tend to have lower female literacy and lower infant mortality than LICs. ☐

f NICs tend to have higher female literacy and lower infant mortality than LICs. ☐

10 Explain why it is important to improve education for girls.

..................................................................................................................................................................

..................................................................................................................................................................

..................................................................................................................................................................

..................................................................................................................................................................

..................................................................................................................................................................

# Theme 8 Environmental challenges

Theme 8 is an optional theme. It is examined in Paper 2, Question 4.

Tick the box if you studied this theme.

## Ecological footprint

Growing food and making consumer goods uses land, resources, and energy. Human activities also create waste. These impacts can be measured as an **ecological footprint**. A footprint of 1 means that each person in a country needs 1 hectare of land to grow and make the things they need.

**1** Match **five** measures to the correct definition below. **Two** terms are not needed.

agri-business   consumerism   ecosystem   ecotourism   e-waste   footprint   food miles

| Term | Definition |
|---|---|
| | The theory that the consumption (use) of goods and services is a good thing because it benefits the economy. |
| | Electronic waste products such as out-of-date computers and mobile phones. |
| | Farming that is organised by large businesses – often by multi-national companies. |
| | Small-scale tourist projects that create money for conservation as well as creating local jobs. |
| | How far food has been transported to get from producer to consumer. |

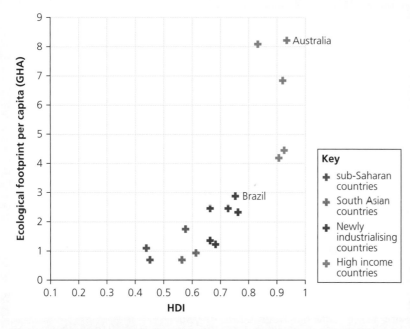

To draw a line of best fit you should try to get an equal number of points on each side. It does not have to go through the origin (where the x-axis and y-axis meet).

**Figure 21 Scatter graph showing the correlation between HDI and ecological footprint**

**2** Add a line of best fit to **Figure 21**.

**3** Study **Figure 21**. Which **three** of the following statements are true? Tick **three** statements.

a There is no correlation between the two sets of data.

b The graph shows a positive correlation.

c In Brazil HDI is 7.5 and ecological footprint is 2.9.

d In Australia HDI is 0.93 and ecological footprint is 8.3.

e NICs have a smaller ecological footprint than sub-Saharan countries.

f HICs have a higher ecological footprint than any other kind of country.

# The impact of consumerism on ecosystems

Food and products are exported across the world to consumers in HICs and NICs. Consumerism has grown due to changes in **technology** and increased **global interdependence**. Consumerism has impacts on the environment and ecosystems including rainforest destruction.

4 Link each point to one explanation using an arrow to explain why consumerism is growing. **One** has been done for you.

| Point | Explanation |
|---|---|
| Mobile technologies and media platforms are changing rapidly... | ...so the population of people who need food, housing and energy is growing. |
| The proportion of people in well-paid jobs in NICs is growing rapidly... | ...so adverts for consumer goods are seen on a growing range of devices. |
| The population of LICs and NICs continues to grow... | ...so there is an emerging group of people who can afford to buy imported consumer goods. |
| Container ships have grown greatly in size in the past 30 years... | ...so perishable goods, like flowers, can be flown further distances. |
| The proportion of people in very highly paid jobs in NICs is growing rapidly... | ...so more consumer goods can be shipped around the world. |
| Aircraft use bigger and more fuel-efficient jet engines... | ...so businesses can make decisions quickly and trade with other people around the world. |
| Undersea cables and satellites have improved communications... | ...so there is a growing number of people who can afford to buy luxury consumer goods. |

**Figure 22 Advantages and disadvantages of growing palm oil in tropical regions such as Borneo.**

| | | |
|---|---|---|
| 1 Palm oil has higher yields than other oilseed crops, such as sunflower. | 2 Palm oil can be grown to make bio-fuel to replace petrol or diesel in cars or machinery. | 3 Animals such as orangutans come into conflict with people who move into the rainforest. |
| 4 Palm oil is a healthier fat alternative for use in food and cooking. | 5 Palm oil needs less fertiliser than other oilseed crops. | 6 The rainforest habitat becomes fragmented so animal movement is restricted. |
| 7 Clearing rainforest and putting in infrastructure, such as roads and electricity supply, provides jobs and incomes. | 8 New roads attract migrants to move into the forest who clear land illegally to make homes. | 9 Palm oil has lower production costs than other oilseed crops, such as soybean and sunflower. |
| 10 Palm oil plantations have far fewer insects than natural rainforest. | 11 Palm oil uses fewer pesticides than other oilseed crops. | 12 The massive biodiversity of the rainforest is replaced by a single species of plant. |

5 Study **Figure 22**. Write the number of each of the twelve statements into the correct place in the following table. **One** has been done for you.

It's not necessary to have the same number of advantages as disadvantages. Some numbers could be used in more than one box.

| | Advantages | Disadvantages |
|---|---|---|
| For Borneo's economy | 1 | |
| For the environment | | |
| For consumers | | |

# Effects of climate change

Climate change has effects on people, the economy and environment. Coral reefs are an example of one environment that is affected by climate change.

6 Link each cause to one effect and one explanation using an arrow to explain why coral reefs are threatened by climate change.

| Cause | Effect | Explanation |
|---|---|---|
| Global warming leads to… | …acidification of sea water | …so the coral structures, which are made of calcium carbonate, dissolve faster. |
| Oceans absorb $CO_2$ which leads to… | …higher ocean temperatures | …so reefs are smothered by sediment and cannot photosynthesise. |
| An increased number of storms leads to… | …rising sea levels | …causing coral bleaching which weakens reef structure. |
| Melting polar ice leads to… | …more soil erosion on land | …so deeper water covers reefs so they receive less sunlight so cannot photosynthesise. |

# Tropical storms are changing

**Climate change** means that the atmosphere is getting warmer. Scientists warn that the most powerful tropical storms will become more common.

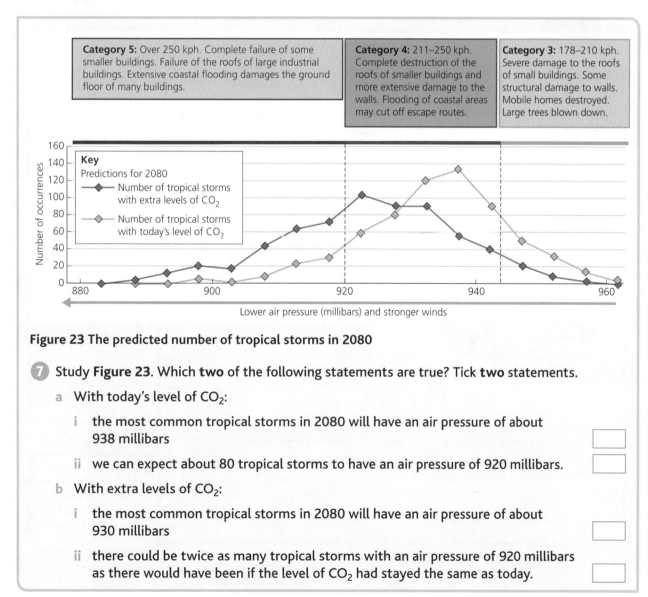

**Category 5:** Over 250 kph. Complete failure of some smaller buildings. Failure of the roofs of large industrial buildings. Extensive coastal flooding damages the ground floor of many buildings.

**Category 4:** 211–250 kph. Complete destruction of the roofs of smaller buildings and more extensive damage to the walls. Flooding of coastal areas may cut off escape routes.

**Category 3:** 178–210 kph. Severe damage to the roofs of small buildings. Some structural damage to walls. Mobile homes destroyed. Large trees blown down.

**Figure 23** The predicted number of tropical storms in 2080

7 Study **Figure 23**. Which **two** of the following statements are true? Tick **two** statements.

a With today's level of $CO_2$:

i the most common tropical storms in 2080 will have an air pressure of about 938 millibars

ii we can expect about 80 tropical storms to have an air pressure of 920 millibars.

b With extra levels of $CO_2$:

i the most common tropical storms in 2080 will have an air pressure of about 930 millibars

ii there could be twice as many tropical storms with an air pressure of 920 millibars as there would have been if the level of $CO_2$ had stayed the same as today.

# Tackling climate change

We can manage climate change using the following approaches:

- **Reducing causes**, for example, by using renewable energy, planting trees to store carbon, or by capturing $CO_2$. This is known as **mitigation**. Politicians from different countries can make **international agreements** about reducing the amount of greenhouse gases that are emitted.
- **Responding to the changing climate**, for example, by changing how we produce food and manage water supplies or by building better flood defences. This is known as **adaptation**.

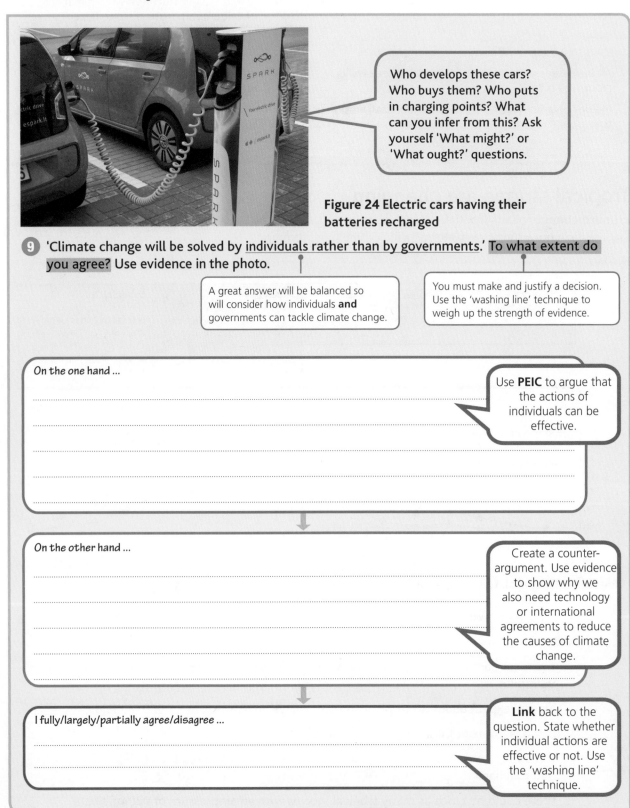

Who develops these cars? Who buys them? Who puts in charging points? What can you infer from this? Ask yourself 'What might?' or 'What ought?' questions.

**Figure 24** Electric cars having their batteries recharged

**9** 'Climate change will be solved by <u>individuals rather than by governments</u>.' To what extent do you agree? Use evidence in the photo.

A great answer will be balanced so will consider how individuals **and** governments can tackle climate change.

You must make and justify a decision. Use the 'washing line' technique to weigh up the strength of evidence.

**On the one hand ...**

.................................................................................................................

.................................................................................................................

.................................................................................................................

.................................................................................................................

.................................................................................................................

Use **PEIC** to argue that the actions of individuals can be effective.

**On the other hand ...**

.................................................................................................................

.................................................................................................................

.................................................................................................................

.................................................................................................................

.................................................................................................................

Create a counter-argument. Use evidence to show why we also need technology or international agreements to reduce the causes of climate change.

**I fully/largely/partially agree/disagree ...**

.................................................................................................................

.................................................................................................................

**Link** back to the question. State whether individual actions are effective or not. Use the 'washing line' technique.

# Managing ecosystems

Wetland ecosystems protect people from floods and help to sustain a clean supply of drinking water. Wetlands are easily damaged by agriculture and urban growth, but they can be **restored**. The London Wetland Centre, in **Figure 25**, is an example of a restored habitat.

**Figure 25** London Wetland Centre

10 Use the photo and phrases to make a **chain of reasoning** that answers the **explain** questions below.

| | | |
|---|---|---|
| store water | release water slowly | store carbon in plants and soil |
| habitat for fish | reduce flooding downstream | help prevent climate change |
| filter pollutants | attract wading birds | improve quality of water |
| supply rivers during drought | create jobs in tourism | |

a Explain why wetlands reduce flood risk.

.........................................................................................................................

.........................................................................................................................

.........................................................................................................................

.........................................................................................................................

b Explain why wetlands should be restored.

.........................................................................................................................

.........................................................................................................................

.........................................................................................................................

# Chapter 4 Assessment of fieldwork

WJEC

## Fieldwork assessment for students in Wales

Students in **Wales** do a fieldwork assessment in November/December of Year 11 called **NEA**. It's an open book test – you are allowed to take your fieldwork portfolio of notes into the assessment. All of the questions are about **your own fieldwork**. There are 40 marks available. Use pages 89–92 to find out more about these kinds of questions.

Eduqas

## Fieldwork assessment for students in England

Students in **England** are examined on their fieldwork in Paper 3 at the end of Year 11. The structure of Paper 3 is shown in **Figure 1**.

**Figure 1 What is assessed in Paper 3 (students in England only)**

| Paper 3 | Applied Fieldwork Enquiry | | Time allowed: 1 hour 30 minutes |
|---|---|---|---|
| **Part** | **Marks** | **What are questions about?** | |
| A | 18 | 1 Fieldwork methods | Answer all questions |
| B | 18 | 2 Fieldwork in which you investigated a concept | Answer all questions |
| C | 36 (+4 for SPaG) | A decision-making exercise based on the same concept as Part B | Answer all questions |

> You should spend about 45 minutes in total on Parts A and B, leaving another 45 minutes for Part C. You could do Part C first. If you do this, make sure you stop after 45 minutes and then go back to Parts A and B.

You can see that Parts A and B are about fieldwork while Part C is a decision-making exercise. There is advice about Part C on pages 92–95. In Parts A and B there are two types of questions about fieldwork:

- Questions about a fieldwork enquiry carried out by someone else. This is called **unfamiliar fieldwork** and you can find some advice about it on pages 86–88.
- Other questions in Parts A and B are about **your own fieldwork**. Use pages 89–92 to find out more about these kinds of questions.

> All questions about fieldwork assess either application or skills. You will never be asked to describe your fieldwork.

### Questions about unfamiliar fieldwork

Questions about **unfamiliar fieldwork** give you some information such as a photograph, table of data, graph or map about fieldwork which has been carried out by someone else. Most of these questions are low tariff (1, 2, 3 or 4 marks). Some important key words are used in fieldwork questions. Make sure you know what they mean:

**accuracy**      **reliability**      **suitable or appropriate**

### Questions which ask about accuracy and reliability

When you collect fieldwork data you want it to be accurate and **reliable**. Data is accurate if the value you have recorded is close to the true value. For example, you could carefully count all of the people in the street shown in **Figure 2** for ten minutes to get accurate data, or you could count for one minute and then multiply by ten to get an estimate. Reliability is about collecting data in consistent ways. For example, imagine five groups of students counted pedestrians in Birmingham. Four groups counted for exactly ten minutes but the fifth group didn't have a watch so counted for what they thought was five minutes. Overall, the data would be **unreliable**.

**Figure 2 A busy shopping street in Central Birmingham**

1 Study **Figure 2**. Suggest why it would be difficult to collect accurate data about the number of pedestrians using this street.

.......................................................................................................

.......................................................................................................

.......................................................................................................

.......................................................................................................

2 A group of students counted the number of pedestrians at eight locations across the city centre. Explain what the students would need to do to make the results reliable.

.......................................................................................................

.......................................................................................................

.......................................................................................................

.......................................................................................................

> Question 1 assesses AO3 – can you analyse the evidence in the photo? The street is busy and people are moving in different directions. The important word in the question is **accurate** – which means the results being as close to the real value as possible.

> Question 2 also assesses AO3 – can you evaluate what the students did? The important word in the question is **reliable**. Reliability is about data being collected in exactly the same way each time.

## Questions which ask about maps and graphs in fieldwork

You may be shown data collected by a group of students presented in a table, graph or map. Some simple questions may ask you a skills question – can you add data to the graph or describe the pattern? Other questions are trickier – you may be asked if the graph/map is suitable or how the graph/map can be improved or adapted. There is more information on this in Chapter 1, page 13.

The students used a bipolar survey (shown in **Figure 3**) to score the quality of the environment at each of the eight locations. The students then chose different techniques to present the bipolar data. These are shown in **Figure 4** and **Figure 5**.

**Figure 3 A bipolar survey**

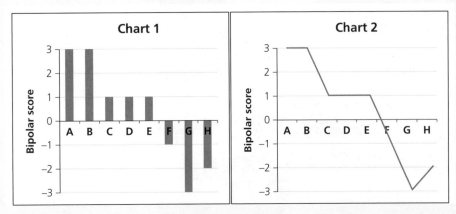

| Positive | +3 | +2 | +1 | 0 | −1 | −2 | −3 | Negative |
|---|---|---|---|---|---|---|---|---|
| Pleasant environment for pedestrians | | | | | | | | Unpleasant environment for pedestrians |

Chart 1

Chart 2

**Figure 4 Different charts representing the bipolar data**

**3** Which is the most appropriate method for presenting the data shown in **Figure 4**? Give a reason for your choice.

.........................................................................................................

.........................................................................................................

The key word here is **appropriate**. Bar charts are appropriate for presenting **discrete data** (things that are counted or put into categories). Line graphs are appropriate for presenting **continuous data** (things that can be measured like height, time or velocity).

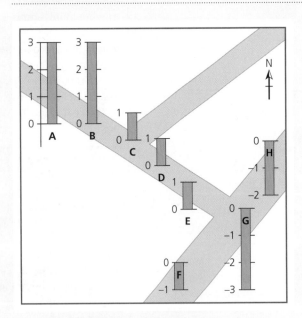

**Figure 5 A map representing the bipolar data**

**4** Describe the pattern of bipolar scores shown in **Figure 5**.

.........................................................................................................

.........................................................................................................

.........................................................................................................

This question assesses geographical skills (AO4). Use the north arrow. Never write about the 'top of the map' or the 'right of the map'.

**5** Suggest one way Figure 5 could be improved.

.........................................................................................................

.........................................................................................................

In this kind of question, always check that the map has a north arrow, key and scale line.

# Questions about your own fieldwork

Students in Wales and England will have some questions that assess **your own fieldwork**. These questions mainly assess your ability to **evaluate** your fieldwork. Use **Figure 6** to make some revision notes. With so many marks available for evaluation it is essential that you also consider the following:

■ **Why** you chose to do each stage of the fieldwork in that way.
■ The **strengths and weaknesses** of each stage of your fieldwork.

When revising you must **think critically** about your fieldwork and **evaluate** what you did and why you did it that way.

**Figure 6 The six strands (or stages) of the fieldwork enquiry**

| Strands | Enquiry 1 | Enquiry 2 |
|---|---|---|
| 1 Choosing suitable enquiry questions | | What was the aim of your fieldwork? Was it sensible and achievable? |
| 2 Selecting, measuring and recording data | | What sampling strategy did you use? Can you justify its choice? Should you have collected any other data? |
| 3 Processing and presenting fieldwork data | | What maps and graphs did you use? What were their strengths and limitations? |
| 4 Describing, analysing and explaining fieldwork data | | Did the data allow you to identify trends and patterns? If not, why not? |
| 5 Reaching conclusions | | Was the data reliable enough so that you could reach firm conclusions? |
| 6 Evaluating your fieldwork | | What might have happened if you had collected your data in a slightly different way or at a different time of day? |

1 Complete **Figure 6** with key details about each of your geography fieldwork enquiries. Use bullet points.

# Evaluation, evaluation, evaluation

You must be able to write a good evaluation of your fieldwork if you want high marks. This isn't easy. Students tend to fall into one of two traps when they are trying to evaluate their fieldwork:

- *Pitfall one*: Some students write about what went well or what went badly, but this isn't necessarily good evaluation. For example, 'I didn't collect much data because it rained' isn't a good evaluation – it means you should have worn a better coat!
- *Pitfall two*: A lot of students make rather vague statements that imply evaluation. For example, 'It was useful to have data from several different sites around the town.' The word 'useful' implies that this was a strength. It would be better to be much more direct. For example, 'It was a **major strength** of my sampling to collect data from several different sites **because** this allowed me to draw a conclusion about the spatial pattern in the data.' This response is much better because 'major' is a qualifying word – it tells the examiner *how* useful the data was. The student then goes on to explain *why* this was useful.

Use words from the lists in **Figure 7** to make sure you are actually signposting your evaluation in a really obvious way. Then the examiner will spot it and, hopefully, give you credit for it.

**Figure 7 Words to use when evaluating your fieldwork**

| Words that describe a positive aspect of your fieldwork | Words that describe a negative aspect of your fieldwork | Qualifying words (adjectives) |
|---|---|---|
| Strength | Limitation | Significant |
| Advantage | Disadvantage | Substantial |
| Benefit | Weakness | Serious |
| Opportunity | Obstacle | Major |
| Merit | Challenge | Minor |
| Success | Failure | Partial |

> Use these words to indicate relative importance, for example, 'minor limitation' or 'major strength'.

# Using your own experience

Lots of students will attempt an evaluation, but it is very limited because it is so general – it could be about any piece of fieldwork. For example, 'We didn't really have a big enough sample.' Adding specific details about **your own experience** is important when you are evaluating your fieldwork. **Figure 8** and **Figure 9** give you some ideas about how to do this.

**Figure 8 How to improve your evaluation by making it specific**

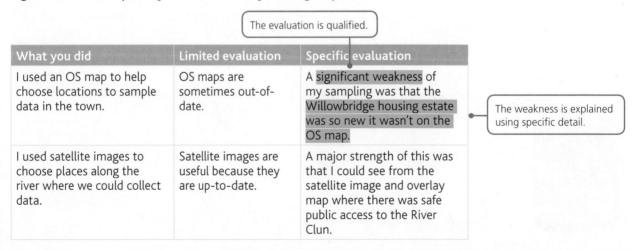

> The evaluation is qualified.

| What you did | Limited evaluation | Specific evaluation |
|---|---|---|
| I used an OS map to help choose locations to sample data in the town. | OS maps are sometimes out-of-date. | A significant weakness of my sampling was that the Willowbridge housing estate was so new it wasn't on the OS map. |
| I used satellite images to choose places along the river where we could collect data. | Satellite images are useful because they are up-to-date. | A major strength of this was that I could see from the satellite image and overlay map where there was safe public access to the River Clun. |

> The weakness is explained using specific detail.

**Figure 9 The difference between limited evaluations and specific evaluations**

> Tick which evaluation is better!

| Data collection | Evaluation | |
|---|---|---|
| I measured the length of pebbles along a transect (a line) going up the beach profile. | The tide was coming in so I had to rush. | ☐ |
| | I didn't check the tide timetable and the tide was coming in. This was a major weakness because I couldn't take pebble sizes at the bottom of the beach profile so my transect is incomplete. | ☐ |
| I intended to use systematic sampling to ask a questionnaire of every tenth person. | Some people didn't want to answer our questions so we just asked people who looked friendly. | ☐ |
| | Some people were too busy to stop and talk so we had to ditch our systematic method and just ask anyone who would stop. This is a problem because our results were probably not representative. | ☐ |
| I measured wind speeds in the sand dunes every twenty minutes. | This was too infrequent because the wind was very gusty on the day we visited because a front had passed over. | ☐ |
| | However, the wind speed varied a lot in between the readings. | ☐ |
| I intended to do an Environmental Quality Index (EQI) every 100 m along a straight line from the suburbs to the city centre. | We couldn't keep to a straight line because there was a main road in the way. This was a minor limitation because it meant our sampling wasn't perfectly systematic. | ☐ |
| | It was difficult to keep to a straight line and do the readings at exactly 100 m. | ☐ |
| I worked as part of a large group collecting data on the amount of traffic all over the town. | One group of students collected their results ten0 minutes later than everyone else so we cannot be sure that the cars they counted hadn't already been counted elsewhere. | ☐ |
| | We had problems getting everyone to collect the results at exactly the same time. | ☐ |

2 a Read each of the evaluations in **Figure 9**. For each example of data collection, tick the evaluation that you think is better.

b Use a pen to circle where the evaluation is qualified.

c Use a highlighter pen to highlight part of the answer that provides **specific details** of the student's own experience.

**3** To what extent were the primary data collection techniques used in either of your fieldwork enquiries effective? [6]

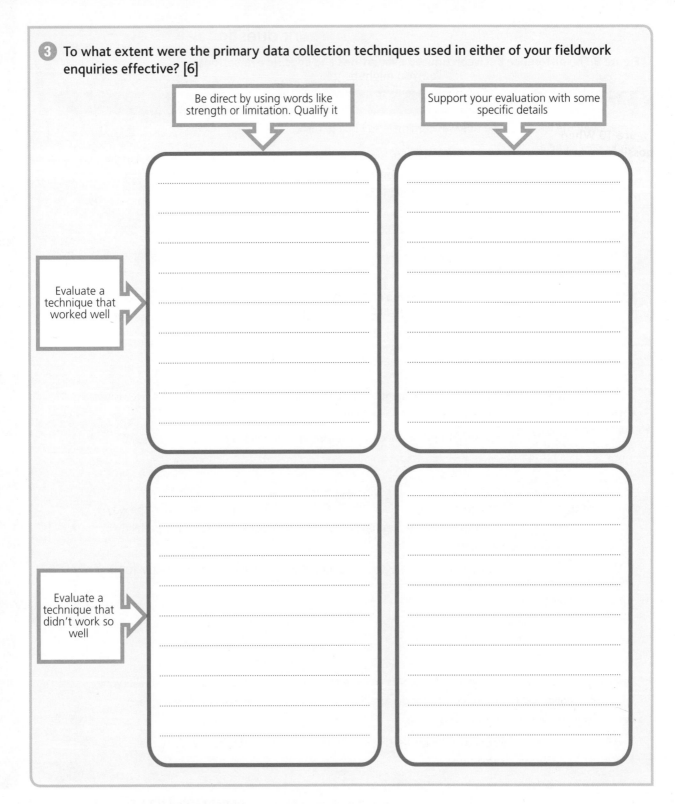

Be direct by using words like strength or limitation. Qualify it

Support your evaluation with some specific details

Evaluate a technique that worked well

Evaluate a technique that didn't work so well

## Preparing for Paper 3 Part C

Eduqas

Part C of Paper 3 is a decision-making exercise. It is worth 36 marks plus 4 marks for spelling, punctuation and grammar (SPaG). You will be asked to study some information about an issue facing a place in the UK. This information could take the form of photos, maps, graphs or Fact Boxes. The questions will assess the following:

■ Your understanding of concepts and processes (AO2). Questions will use command words such as *Give one reason* or *Explain why*. There is advice on this type of question on pages 9–10.

■ Your geographical skill in dealing with this evidence (AO4). There is advice on this type of question on pages 12–17.

Part C finishes with a big question worth 12 marks (plus 4 marks for SPaG). This question will ask you to make your decision and justify it.

To justify means to give the reasons for a decision.

# Planning your answer to the Part C, 12-mark question

There are a total of 16 marks (12 plus 4 for SPaG) available, and you have about fifteen minutes to complete your answer. You might be tempted to start writing straightaway, but you ought to plan your answer first. You could use a mind-map or a simple table like the one shown in **Figure 10**.

**Figure 10 When you are planning your answer, use a table to evaluate the possible impacts of your decision**

|  | Advantages of my decision | Disadvantages of my decision |
|---|---|---|
| SOCIAL (for people and communities) | | |
| ENVIRONMENTAL (water use, ecosystems, air pollution) | | |
| ECONOMIC (jobs and businesses) | | |

## Structuring your 12-mark answer

**Figure 11** shows how you might structure your answer in three simple paragraphs.

> Two minutes writing a plan is time well spent. Keep glancing back at your plan – and the question – as you write your full answer.

Paragraph 1
**I think the best option is ...**

> Use the **PEIC** technique to build an argument that supports your decision.

Paragraph 2
**I have rejected the other options because ...**

> Use the **PEIC** technique to explain why you have rejected the other options.

Paragraph 3
**In conclusion ...**

> Conclude by reminding the examiner of the most persuasive pieces of evidence.

**Figure 11 How to structure your decision**

It is very important that you explain why your decision is best. Do this in the first paragraph. You might realise that your decision could have some negative implications. That's okay, but it's important to explain why the advantages of your plan outweigh any disadvantages.

In the second paragraph, explain why the other options are not as good as the one you have chosen. It's a good idea to give a balanced account, so you should write about the possible advantages of these options before explaining why, on the whole, you have rejected them.

Finish your answer with a short conclusion. **Figure 12** gives you some advice about this final paragraph.

> If you use the word 'sustainable' make sure you explain **why** your option is sustainable.

**Figure 12 Dos and don'ts of writing a conclusion**

| Do | Don't |
|---|---|
| Do remind the examiner that you have looked at both sides of the argument. | Don't sit on the fence! If you have been asked to make a judgement, you should make it. |
| Do repeat what you think is the most persuasive or significant piece of evidence that supports your decision. | Don't state that it is difficult to make a decision if the earlier part of your essay is packed full of strong evidence. |
| Do use key words or phrases from the question in your conclusion. | Don't forget to glance back at the question before you start the conclusion. If you have wandered off task then now is your chance to save your essay and answer the question! |

## Signposting

Examiners like the 12-mark answers to be structured so a good answer will be organised into paragraphs. You can also use signposting to help structure your answer. Signposting is a technique that tells the reader what is coming next – like a signpost on the roadside tells you where you are going. **Figure 13** has a few useful signposts you can use when writing your longer answers.

**Figure 13 Useful signposts**

| Adding points | Writing in lists | Adding emphasis | Linking cause and effect |
|---|---|---|---|
| Moreover | Next | Especially | As a result |
| Furthermore | Then | Particularly | Because |
| In addition | Lastly | Chiefly | Consequently |
| What is more | Firstly | Mainly | Despite this |
| | Secondly | Mostly | Therefore |
| | Finally | | |

## Aiming high!

Your decision will have impacts and your answer must show that you understand this. Evaluating the possible social, economic and environmental impacts is good. However, if you are aiming for a really high mark then you should go beyond this and evaluate one or more of the following:

- Possible consequences of your decision on **different** groups of people.
- Likely **spatial** impacts of your decision – local, regional or national.
- How your plan could have **short-term** and **long-term impacts**.

To help with this, think about the consequences of the damage to the railway track in **Figure 14**. A good candidate might use this resource to suggest that, until the line is repaired, commuters will have to use a bus service for part of their journey and this would be a significant inconvenience. They might also suggest that the consequences were local (repairs to the sea wall to make local properties safe) and national (Network Rail had to spend millions to repair the track – money that could not be spent elsewhere).

When suggesting possible impacts you should avoid extreme statements. For example, using **Figure 14**, it would be extreme to state that commuters wouldn't be able to get to work and that they would lose their jobs!

**Figure 14 Storm damage to the railway line at Dawlish in Devon**

**3** Study **Figure 14**. This railway line joins Cornwall to London. It was damaged in February 2014. It took two months and £35 million to make the repairs. Outline the possible impacts on each of the following:

**a** Three different groups of people.

.........................................................................................................................

.........................................................................................................................

.........................................................................................................................

.........................................................................................................................

.........................................................................................................................

**b** The local area (Dawlish) and the region of Cornwall.

> Think about impacts on local residents and businesses in Cornwall.

.........................................................................................................................

.........................................................................................................................

.........................................................................................................................

.........................................................................................................................

.........................................................................................................................

**c** One short-term (during February-March 2014) impact and one longer-term (after April 2014) impact.

> Think about how long it might take business to recover lost trade.

.........................................................................................................................

.........................................................................................................................

.........................................................................................................................

.........................................................................................................................

.........................................................................................................................

**①** **This workbook will help you** prepare for your WJEC GCSE Geography and Eduqas GCSE (9–1) Geography A exam.

**②** **Build your skills** and prepare for every question in the exam using:
- clear explanations of what each question requires
- short answer activities that build up to exam-style questions
- spaces for you to write or plan your answers.

**③** **Answering the questions** will help you build your skills and meet the assessment objectives (AOs):

**AO1:** remembering geographical facts

**AO2:** understanding geographical concepts and processes

**AO3:** evaluating evidence or using evidence to make a decision

**AO4:** using skills to investigate maps and graphs or making calculations.

**④** **You still need to** read your textbook and refer to your revision guide and lesson notes.

**⑤** **Answers** to every question in the book are available at **www.hoddereducation.co.uk/workbookanswers**

The Publishers would like to thank the following for permission to reproduce copyright material.

**Photo credits**

p. 18 © Robert Douglas / Alamy Stock Photo; p. 26 t © Commission Air / Alamy Stock Photo; b © A.P.S. (UK) / Alamy Stock Photo; pp. 28, 29, 30, 32 © Andy Owen; p. 33 © Alan Curtis / LGPL / Alamy Stock Photo; p. 35 © John Potter / Alamy Stock Photo; p. 40 © Andy Owen; p. 44 © RichSTOCK / Alamy Stock Photo; p. 48 b © knlml - stock.adobe.com; p. 52 © migstock / Alamy Stock Photo; p. 54 t and b © Andy Owen; p. 55 © incamerastock / Alamy Stock Photo; p. 57 © Andy Owen; p. 62 © Rainier Martin Ampongan / Alamy Stock Photo; p. 65 © Andy Owen; p. 66 © Hemis / Alamy Stock Photo; p. 71 © Andy Owen; p. 72 © Agencja Fotograficzna Caro / Alamy Stock Photo; p. 75 © Andy Owen; p. 79 © Sk Hasan Ali/Shutterstock.com; p. 84 © Shutterstock / Anton Gvozdikov; pp. 85, 87 © Andy Owen p. 95 © Lightworks Media / Alamy Stock Photo.

**Acknowledgements**

Map on p.14 © Crown copyright and database rights, 2019, under licence to Ordnance Survey. Licence number 100036470.

Every effort has been made to trace all copyright holders, but if any have been inadvertently overlooked, the Publishers will be pleased to make the necessary arrangements at the first opportunity.

Although every effort has been made to ensure that website addresses are correct at time of going to press, Hodder Education cannot be held responsible for the content of any website mentioned in this book. It is sometimes possible to find a relocated web page by typing in the address of the home page for a website in the URL window of your browser.

Hachette UK's policy is to use papers that are natural, renewable and recyclable products and made from wood grown in well managed forests and other controlled sources. The logging and manufacturing processes are expected to conform to the environmental regulations of the country of origin.

Orders: please contact Hachette UK Distribution, Hely Hutchinson Centre, Milton Road, Didcot, Oxfordshire, OX11 7HH. Telephone: +44 (0)1235 827827. Email education@hachette.co.uk Lines are open from 9 a.m. to 5 p.m., Monday to Friday. You can also order through our website: www.hoddereducation.com

ISBN: 978 1 5104 5351 7

© Andy Owen 2019

First published in 2019 by
Hodder Education,
An Hachette UK Company
Carmelite House
50 Victoria Embankment
London EC4Y 0DZ

www.hoddereducation.co.uk

Impression number 10  9  8

Year        2024

Cover photo © Aurora Photos/Alamy

Illustrations by Aptara Inc.

Typeset in India by Aptara Inc.

Printed in India

A catalogue record for this title is available from the British Library.

**HODDER EDUCATION**

t: 01235 827827
e: education@hachette.co.uk
w: hoddereducation.co.uk

ISBN 978-1-5104-5351-7

9 781510 453517

MIX
Paper | Supporting
responsible forestry
FSC™ C104740